SAMPLING METHODOLOGIES
WITH APPLICATIONS

CHAPMAN & HALL/CRC
Texts in Statistical Science Series

Series Editors
C. Chatfield, *University of Bath, UK*
J. Zidek, *University of British Columbia, Canada*

SAMPLING METHODOLOGIES
WITH APPLICATIONS

Poduri S.R.S. Rao

Professor of Statistics
University of Rochester
Rochester, New York

CHAPMAN & HALL/CRC

Boca Raton   London   New York   Washington, D.C.

Library of Congress Cataloging-in-Publication Data

Rao, Poduri S.R.S.
  Sampling methodologies : with applications / Poduri S.R.S. Rao.
    p.   cm. --(Texts in statistical science)
  Includes bibliographical references and index.
  ISBN 1-58488-214-X (alk. paper)
    1. Social sciences—Statistical methods. 2. Sampling (Statistics) I. Title. II. Series.

HA31.2 .M555 2000
519.5'2—dc21
                                                                    00-063926

Contents

List of tables

List of figures

To Durgi, Ann, and Pal

Preface

This book contains the methodologies and applications of commonly used procedures of sampling from finite populations. Easy access of the different topics to students and their quick availability to practitioners have been the objectives in preparing this book. With these objectives, the main results along with their logical explanations are presented in the text, but the related derivations are placed in the appendixes to the chapters.

The introductory chapter describes the differences between probability and nonprobability sampling, and presents illustrations of several types of national and international surveys. Properties of simple random sampling along with the estimation of the population mean, total, and variance are considered in the second chapter. Unbiasedness of the sample mean, its variance, and standard error are defined and illustrated. Sample sizes needed to estimate the population mean or total with specified criteria are also examined in this chapter.

The third chapter contains a variety of topics of importance in sample surveys and for statistical analysis in general. Definitions of the bias, variance, and mean square error of an estimator and their relevance to its precision and accuracy are presented and illustrated first. Further topics include the covariance and correlation between two random variables and their sample means, tests of hypotheses related to the means, and the comparison of simple random and systematic random sampling procedures. The effect of the bias arising from the nonresponse in a survey is briefly examined in this chapter, and this topic is continued in the exercises of some of the following chapters and finally in detail in Chapter 11. The appendix to this chapter contains the basic results on conditional and unconditional expectation and variance, and other topics of importance.

Chapters 4 through 11 contain detailed presentations of the estimation of percentages and counts, stratification, subpopulations, single and two-stage cluster sampling with equal and unequal probabilities, ratio and regression methods of estimation, and the problem of nonresponse and its remedies. The final chapter presents in detail linearization, jackknife, bootstrap, and balanced repeated replication procedures. Major results on small-area estimation and complex surveys appear in this chapter.

Each topic is presented with illustrations, followed by examples and exercises. They are constructed from data on everyday practical situations covering a wide variety of subjects ranging from scholastic aptitude tests to health care expenditures and presidential elections. The examples and exercises are interwoven throughout different chapters.

For the sake of good comprehension of the results on unbiasedness, standard error, and mean square error of the estimators, some of the exercises are constructed as projects that require the selection of all the possible samples of a specified size from a finite population. For some of the exercises and projects, Minitab, Excel, and similar computer programs are quite adequate. The methodological type of questions at the end of the exercises along with the derivations in the appendixes should be of particular interest to advanced students. Solutions to the odd-numbered exercises are presented at the end of the book. Detailed solutions to all the exercises appear in the *Teacher's Manual*.

This book can be recommended as a text for a one-semester or two-quarter course for students in statistics and also in business, political and social sciences, and other areas. One or two courses in basic theoretical and applied statistical concepts would provide the required preparation. This book can also serve as a reference guide for survey practitioners, political and public pollsters, and researchers in some industries.

I am grateful to Dr. Ramana L. Rao for patiently assembling the entire manuscript on the word processor. I would like to thank Bob Stern for his interest in this project and his editorial suggestions. Thanks to Megan Brooks, Sharon Gresham, and Chris Petrosino for checking the solutions to some of the exercises and their suggestions for presenting the solutions to the examples. I would also like to thank my wife Durgi and children Ann and Pal for their encouragement and patience throughout the years spent preparing this manuscript.

Poduri S.R.S. Rao
University of Rochester, New York

CHAPTER 1

Introduction

1.1 Introduction

Sample surveys are conducted throughout the world on topics related to agriculture, demography, economics, education, health and welfare, politics, and several other topics of interest. National and international organizations and private agencies collect information through sample surveys on items such as agricultural and industrial productions, employment rates, educational levels, and medical services. Information on the public is also collected on demographic items such as age, marital status, and family size and on socioeconomic characteristics such as religion, ethnicity, and occupation.

Information from these surveys is essential, for instance, for agricultural and economic planning, for making improvements in educational and medical programs, and for providing public services. Estimates and predictions of characteristics such as the number of students graduating from colleges and their starting salaries, information on medical services needed for the young and the old, and opinions of the public on political issues are obtained from data collected through surveys.

In the U.S., the Departments of Agriculture, Commerce, Education, Health, and Labor and other government agencies regularly conduct large-scale surveys on the population. *Statistics Canada*, the *National Sample Surveys* in India, and the United Kingdom routinely conduct agricultural, demographic, and economic surveys. Similar surveys are also conducted at regular intervals in several countries in Africa, Asia, Australia, Europe, and Latin America.

At the international level, the Food and Agricultural Organization (FAO), the World Health Organization (WHO), and other agencies of the United Nations conduct surveys throughout the world. The National Household Survey Capability Program of the UN is a cooperative undertaking in Africa and other countries of the world. Collection and dissemination of data related to agricultural productions, economic activities, and the demographic and socioeconomic aspects of the populations of these countries are some of the major purposes of this program. The World Fertility Survey (WFS) was conducted by the UN between 1972 and 1984 to collect information on the birth rates, family sizes, and other related characteristics from the households in more than 60 countries around the world.

Offices of the presidents, prime ministers, cabinet members, and legislatures employ sample surveys for ascertaining public opinion on impending legislations and reforms. Political scientists use survey data to study the opinions and attitudes of people on the candidates for public offices, defense budgets, government spending, issues related to war and peace and the like. Exit polls are used for predicting the outcomes of political elections. Opinions on school budgets, library facilities, recreational activities, and other public services in towns and cities are frequently obtained from samples of automobile or voter registrations.

Social scientists analyze information from surveys to examine the living conditions, changes in educational standards and employment rates, and the general welfare of the people. Major newspaper, television, and radio networks throughout the world summarize the information collected from polls and surveys on current topics of interest. During times of earthquakes, epidemics, famines, floods, wars and similar exigencies, sample surveys are usually the only sources for obtaining quick information on the affected public.

Surveys are conducted by market research firms to assess the preferences of consumers regarding the household purchases, public services, and opinions on the quality and prices of consumer products. Samples are routinely used for industrial quality control. Performances and reliabilities of automobiles, televisions, refrigerators, computers, and stereo equipment are examined through selected samples.

Television ratings are based on the information collected from a sample of households. The Dow–Jones Industrial Average is computed from a sample of only 30 from almost 3000 stocks. The S&P 500 index is obtained from a sample of 500 stocks selected from specified groups of the stocks. Similar indices are obtained from samples of particular types of stocks. Sample surveys are also used for auditing financial accounts.

Sampling is frequently employed to estimate the sizes of fish, bird, and animal populations and the effects of environmental pollution and ecological changes on their survival. There are many more uses of sample surveys. Information obtained from properly selected samples and the accompanying statistical estimation and inference procedures can be vital in the daily activities of the people. The essential aspects of sample surveys are described in the following sections.

1.2   Censuses and surveys

The Decennial Census on the population and housing in the U.S. is conducted by the Bureau of the Census. In this type of census, which is also conducted in several other countries, it is required to count every person in the country and obtain information from all the households on a number of demographic and socioeconomic characteristics. In addition to the population and housing censuses, other types such as the censuses of manufacturing industries and medical services are conducted periodically in many countries. Considerable amounts of money and time are expended for collecting the required information through the censuses and for analyzing it. It was reported that the 1990 census of population on approximately 100 million households in the U.S. was conducted at a cost of more than \$2.6 billion and almost a year was needed to assemble the required information.

In the context of sample surveys, a collection or an aggregate of units such as people, households, cities, districts, countries, states, or provinces is called a **finite population**. A sample consists of only a portion of the population units. A census is a 100% sample and it is a **complete count** of the population.

To collect information on the population characteristics, there are distinct advantages in considering a sample instead of a census. First, a sample provides timely information with relatively low cost. Second, a number of errors can occur in a census for collecting, recording, tabulating, and analyzing the vast amount of information. In a carefully conducted survey, such mistakes can be avoided to a large extent. Information on employment rates, educational levels, and other characteristics of the population of national interest is also needed between the census periods. The national and international surveys mentioned in the previous section provide such information. Sample surveys are also conducted during a census to obtain information on variables not included in the census.

Both censuses and sample surveys may fail to include some of the population units, and may not be able to contact some of the included units. Even if they are contacted, it may also become difficult to obtain the required information from some of the contacted units. These types of noncoverage, noncontact, and nonresponse are known as the

nonsampling errors. These errors and errors of observations, measurements, and recording occur in large-scale censuses more frequently than in sample surveys.

1.3   Types of surveys

Many of the national and international surveys are usually of the **multisubject** and **multipurpose** type. For example, a household survey may include a number of demographic and socioeconomic characteristics such as household sizes, incomes of the heads of households, number of children attending elementary schools, and family finances.

In almost all small and large-scale surveys, before selecting the samples, the population is first divided into a fixed number of **strata** consisting, for example, of geographical regions or groups of units defined according to characteristics such as age, education, income level, sex, and race. Samples are selected independently from each stratum. For many types of demographic and economic surveys, the strata consist of states or provinces and cities. For agricultural surveys, strata are usually formed by dividing the entire land into regions or zones. In some cases, each stratum is further divided into substrata; for example, rural and urban areas of a state.

For large-scale surveys, well-defined **clusters**, groups of contiguous units, may be considered in each stratum. Counties in states, districts in provinces, enumeration areas in rural zones are some examples of clusters. Households, individuals, corporations, and farms may form the final sampling units from which the needed information is obtained.

The actual procedure of sampling in some situations may be implemented in two, three, or more stages. In a household survey, at the first stage, a sample of counties may be drawn from a stratum. At the second stage, a sample of blocks or areas may be drawn from each of the selected counties. At the third stage, a sample of dwellings may be drawn from each of the selected blocks. **Multistage sampling** of this type is a common feature of large-scale surveys.

Several surveys conducted by the government and private organizations are repeated periodically. In the **rotation** or successive sampling procedures, observations are made on some of the units selected previously and on some new units sampled at the current period. In **longitudinal** surveys, observations are made on a selected sample, a **panel**, over periods of time.

For any of the above types of surveys, the population units may be selected into the sample with **equal** or **specified** probabilities, **systematically** or through some other suitable procedure. These approaches are presented in detail in the following chapters.

1.4   Sampling frame

The sample is selected from the **frame**, the list of all the units of the population to be surveyed. The frame should contain all the units of the population under consideration, the **target population**. For example, to estimate the educational expenses of students in a college or university, a sample from the frame with only the names of students living in the dormitories and residence halls will be inadequate. It should also include the names of the students living off-campus. The residents of a metropolitan area may be listed in more than one frame; for instance, automobile registrations and city directories. If a population consists of groups or strata, separate lists of the units in each stratum should be available. Similarly, for the multistage procedures, detailed lists for each stage of sampling should be available.

To estimate agricultural production in some countries, the total region under cultivation is divided into **enumeration districts** (EDs) and **enumeration areas** (EAs) and listed in a frame. For the *aerial* surveys conducted through satellite images, the frame is constructed by dividing the map of the area under consideration into square or rectangular grids.

1.5   Questionnaires, interviews, and sample sizes

Questionnaires

In almost every major survey, information on more than one characteristic is solicited from the respondents. The questionnaire should contain all the items in the form of questions to be answered by the respondents or statements to be completed. "How many persons are in your household?" and "How much do you spend each week on (a) food and (b) entertainment?" are examples of the quantitative type of questions. On attitude and opinion type of questions, the answers to be checked by the respondents usually are in the form of *Yes, No,* or *Don't Know*. To record the tastes, likes, and dislikes on the specified characteristics in a survey, respondents are sometimes asked to express their preferences on a scale ranging from 0 to 5 or 10.

For quite a long time, employment rates were estimated by the U.S. Department of Labor by asking the adult male respondent whether he was employed at the time of the interview, adult female whether she was "keeping house," and the teenage respondent whether he or she was attending school. This method of questioning was found to be underestimating the unemployment rate, especially of women staying home and seeking employment. Starting in January 1994, it was decided to estimate these rates by asking questions with Yes–No answers.

Another illustration that emphasizes the importance of correct words and sentences in a questionnaire is provided by a survey on the daylight saving time in the U.S. Clocks are set back an hour in the fall and advanced an hour in the spring. This practice provides an extra hour of morning sunlight in the fall. In *The New York Times* of October 30, 1993 a reader cited a 1981 survey in which two thirds of the American public favored an extra hour of morning sunlight throughout the year. Another reader responded that the results of such a survey would probably be different if the public were asked whether it would prefer an extra hour of sunlight in the morning or afternoon.

Interviewing methods

Information from the sample units can be obtained through (1) mail, (2) telephone, and (3) face-to-face personal interviews. In several countries, postal and telephone services are extensively used for surveys. Personal interviews are usually the only means of conducting surveys in the rural areas of many countries. Computer aided telephone interviewing (CATI), in which the interviewer equipped with a headphone attempts to obtain responses, is practiced in industrialized nations. During and after a debate, discussion, or presentation on topics of public interest, soliciting responses through the Internet, e-mail or directly on the television is rapidly becoming popular.

Sample sizes

For a survey on the families in a region, the number of households to be selected in the sample has to be determined before the start of the survey. For a multistage survey, sample sizes for all the stages should be specified in advance. **Precision** of an estimator, which is defined in Chapter 2, will be high if the variation among all the possible estimates that can be obtained from a probability sampling procedure is small. To determine the size for a planned sample, importance should be given to the required precision, error of estimation, costs of sampling, and the financial resources as well as the time available for the survey.

1.6   Probability sampling

All the sampling procedures described in the following chapters are of this type. Samples selected through probability sampling are also frequently known as **random samples**. Randomness, however, does not imply that the sample units are selected haphazardly. On the contrary, in this method of sampling, the probabilities for selecting the different samples from a population are specified, and the probabilities for the units of the population to appear in the samples are known.

   Sampling error is a measure of the departure of all the possible estimates of a probability sampling procedure from the population quantity being estimated. A very important feature of **probability sampling** is that in addition to providing an estimate of the unknown population quantity, it enables the assessment of the sampling error of the estimate, the **standard error**. Further, the estimate and its standard error can be used to obtain **confidence limits** for the unknown population quantity. For a specified probability, these limits provide an interval for the population quantity. Improved procedures of probability sampling to suit practical situations and to estimate population quantities with the desirable properties of high **precision and accuracy** will be presented throughout the following chapters.

1.7   Nonprobability sampling

Purposive sampling

To estimate the total sales or average prices of computers in a metropolitan area, a handful of selected computer stores may thought to be adequate. To estimate the agricultural production in a village, a sample consisting of a certain number of small, medium, and large farms may thought to be **representative**. Similarly, some specified households may be selected for a survey on the family finances in a region. In these illustrations, prior information on the population units is frequently utilized for selecting the sample.

Quota sampling

In this method, the survey is continued until a predetermined number of the people, households, hospitals, corporations, and similar population units with specified characteristics are contacted and interviewed. In a political survey, for example, the interviewers may be asked to obtain 200 responses from each of the male and female groups between the ages of 25 and 65. In several surveys, people are contacted until specified percentages of responses are obtained to the different items in the questionnaires.

Other types of surveys

In what are known as marketplace mall surveys, interviewers attempt to elicit responses from the shoppers regarding their preferences for consumer products and political candidates, and the results of these surveys are projected to the general public. Fashion magazines elicit responses to questions on the tastes, preferences, attitudes, and opinions of the readers on various topics of interest and extrapolate the results to the entire population.

In some situations, without a proper sampling procedure, it may become convenient to obtain the required information from the readily available population units. Estimates and predictions from these types of nonprobability sampling procedures can sometimes be close to the population values, but their success cannot be guaranteed. Further, the errors of these estimates and predictions cannot be estimated from these types of samples, unlike in the case of probability sampling procedures.

1.8   Sampling in practice

Different procedures of probability sampling are employed suitable to the practical situations. Some of the typical surveys and procedures conducted in practice are presented in the following subsections with brief descriptions of the terminology and procedures of probability sampling. Illustrations of these procedures will be presented throughout the coming chapters.

Demographic and economic surveys

The U.S. Bureau of the Census conducts the Decennial Census of Population and Housing and also a number of ongoing demographic, economic, and household surveys. The bureau also provides services to other government agencies and conducts research on various statistical procedures related to censuses and surveys.

For the Current Population Survey (CPS), conducted every month by the above organization, primary sampling units (PSUs) are first constructed from counties or groups of counties. The PSUs are then grouped into strata with similar characteristics and of approximately equal sizes. At the final stage, about four housing units, the ultimate sampling units (USUs) are selected. For some of the household surveys, the CPS follows a **rotation sampling** scheme. In this procedure, the households selected in the initial sample are included in the sample for 4 months, left out for 8 months, and included again for a further 4 months.

Educational surveys

In several countries, surveys are conducted to examine the educational programs in schools and colleges. Griffith and Frase (1995) describe some of the **longitudinal surveys** of the U.S. Department of Education, which are repeated over periods of time. Assessing the progress and achievements in the schools and colleges and examining the "quality, equity and diversity of opportunity" are the major objectives of these studies. A sample of 8000 students was selected in 1989–90 for the longitudinal study on students completing high school, and a sample of 11,000 was selected for the study on students completing college. The probability samples were selected in two stages. Schools were selected first and then students were chosen from the selected schools.

Employment and expenditure surveys

During the four quarters in a year, the U.S. Bureau of Labor Statistics conducts surveys on the employment and unemployment rates of people, labor conditions, wages, industrial production, technological growth, consumer expenditures, and related topics.

Jacobs et al. (1989) describe the Consumer Expenditure Survey conducted by this bureau. Two ongoing surveys are conducted to obtain information on family expenditures and living conditions and for

providing information for the **Consumer Price Index**. The entire
nation is divided into 104 geographical areas, 88 urban and 16 rural,
and probability samples are selected from each area. In the Interview
Survey, a sample of about 9000 families are contacted and approximately
5000 completed interviews are obtained. The responses are elicited on
expensive items like property, rent, automobiles, appliances, utility bills,
and insurance premiums. The Diary Surveys are completed at home by
a sample of 8000 families on characteristics difficult to recall, such as
the expenditures for food, medicine, household supplies, and services.

Health and nutrition surveys

In the U.S., the National Center for Health Statistics conducts a number
of surveys on the health and nutritional conditions of inhabitants.
Massey et al. (1989) described these surveys for collecting information
on the number of visits to hospitals and physicians, short-term and
long-term disabilities, chronic conditions, and related characteristics.
Approximately, 132,000 persons from a sample of 49,000 households
are personally interviewed to obtain the required information.

For these surveys, the entire country is divided into 1900 PSUs. A
PSU consists of a county, a small number of contiguous counties, or a
metropolitan statistical area (MSA). These PSUs are grouped, **stratified**, according to demographic and socioeconomic variables such as age,
sex, race, education, income, and the like. Each PSU is divided into area
segments, and each segment is divided into clusters of about eight
households. The country is divided into approximately 800 clusters. The
actual sampling is conducted in two stages. At the first stage, a sample
of PSUs is selected from each stratum with probability proportional to
its size. The largest PSUs are included with certainty. At the second
stage, a sample of clusters is selected and the households in the selected
clusters are interviewed.

Brackstone (1998) describes the **longitudinal survey** to determine
the factors affecting Canadian children's physical, social, and cognitive
characteristics. A sample of 23,000 children from birth to 11 years of
age is selected from 13,500 households. Measurements on the above
characteristics are obtained every 2 years on each child until age 25.
"In each wave, a new sample of 5000 children" from birth to 23 months
of age are added to the sample.

Sampling procedures for the World Fertility Survey was described
by Verma et al. (1980). Demographic and health surveys conducted in
the developing countries through the cooperation of the UN and WHO

are described by Verma and Thanh (1995). For these surveys, the sampling area in each country is stratified according to the urban or rural type and by the location. At the first stage of sampling, a sample of area units is selected with probability proportional to their sizes. At the second stage, a sample of households is chosen from the selected area units. From women between 15 and 49 years of age in the chosen households, information is obtained on their ages at marriage, reproductive history, number of children, immunization of the children, and other demographic and health-related characteristics.

The Health and Nutrition Survey in Croatia on children 5 years old and younger, conducted in 1996 by the UN International Children's Educational Foundation (UNICEF) was described by Dumicic/ and Dumicic/ (1999). The region unaffected by the war, "front line areas," and "liberated areas" formed the three strata. Each stratum was divided into area segments. In each segment, clusters of 40 households with children under 5 years old were considered. The segments consisted of 120 to 250 households. At the first stage, samples of segments were selected from each stratum with probabilities proportional to the sizes of the segments. At the next stage, samples of clusters were selected from the chosen segments. This procedure resulted in the selection of a sample of 14,800 households consisting a total of approximately 2000 children under age 5.

Edler et al. (1999) describe the Heidelberg Children Health Survey conducted in 1996 on 3828 fifth-grade students. It was repeated in 1998 on 4036 seventh-grade students in 172 classes in 68 schools of the lower, middle, and higher levels of education. The questionnaire with 38 items related to the health of young people was presented to the selected students. The above authors presented data on the smoking habits of the students and their parents, and the frequency of their resulting headaches. It was found that children in families where one or both parents smoke suffered from headaches more frequently than nonsmoking families. There were also significant differences in the smoking habits of children in the above three types of schools.

Agricultural surveys

These surveys are conducted to estimate the acreage and production of crops, livestock numbers, expenditures for agricultural production and labor, utilization of fertilizers and pesticides, and related topics.

Kott (1990) summarizes six surveys conducted by the U.S. Department of Agriculture during 1989–90. In each of the 50 states, farms of similar sizes are grouped into strata. Samples of farms are selected from the frames, the lists of the farms in the strata. Samples are selected from lists available separately for crops, farm animals, and farm labor. Area frames are also used for the above surveys. In this procedure, each state is divided into segments and a sample of segments is selected. The required information is obtained from the tracts of farm land utilized for agricultural production.

For the agricultural surveys in India described by Narain and Srivastava (1995), each state is stratified into administrative regions. At the first stage, a sample of villages is selected from each region. At the second stage, a sample of agricultural holdings is chosen from each of the selected villages and the required information is obtained.

Implementation of the surveys in Lesotho and Sudan were described by O'Muircheartaigh (1977). Lesotho is a small mountainous country with many rivers and streams, and the ecological zones and population densities were used for stratification. For the agricultural surveys, administrative districts, ecological zones, and the number of cultivated fields were used for stratification. Sudan is a very large and thinly populated country with a vast amount of desert land. The administrative structure within the provinces, rural councils, and such are used for implementing the surveys.

For the agricultural surveys in several countries, each cultivated region is stratified into blocks. At the first stage, a sample of villages is selected from each block. At the second stage, a sample of land holdings is chosen from each of the selected villages and the required information is collected. For the multistage surveys conducted by the UN in Ghana, Kenya, Nigeria, and other African countries to obtain information on agricultural, demographic, and economic characteristics, EAs are selected at the first stage. Landholdings or dwellings are selected at the second stage.

Carfagna (1997) reviews surveys in the European Union conducted through high-resolution satellite images. The entire area is divided into square **sites**, and each site is divided into square segments. A sample of sites is selected first, and then a sample of **segments** is chosen from each of the selected sites. At the next stage, a sample of **points** is chosen from each selected segment and the different crops are estimated.

Marketing surveys

Almost every major company that manufactures or sells consumer products or provides services to the public conducts surveys on the satisfaction and opinions of the consumers. Dutka and Frankel (1995) describe market surveys in 19 countries on persons 12 to 49 years of age on the brand-awareness of soft drinks, their attitudes toward the brands, and consumption habits. Area sampling is used for these surveys. Each region is **stratified** into urban, rural, metropolitan, and similar divisions. Each stratum consists of PSUs—counties, cities, and similar units. At the first stage, a sample of PSUs is selected. At the second stage, a sample of blocks is chosen from each of the selected PSUs. In the households of the chosen blocks, interviews are attempted on ten people, five males and five females.

Surveys on retail stores in Latin America conducted by the Nielson Marketing Research were described by Santos (1995). The purpose of these surveys was to measure the activity in the stores such as counting of the stocks and purchases, recording the prices, and promotional activities. For these surveys, stratification was based on the geographical region, size of the city, and type of the store, for example, pharmacies and kiosks, and also by the size of the stores determined by the number of employees, selling area, and related information.

Election surveys

In every democratic country, surveys and polls are conducted on the opinions of the people regarding the candidates contesting for public offices such as the president, governors, senators, representatives, and mayors, and the issues of importance at the time of the elections. They are also conducted to forecast the election outcomes ahead of the complete counts of the ballots. In the U.S., television and newspaper networks and several national and local polling organizations conduct such surveys. Brown et al. (1999) describe the procedures followed for the forecasts for the four political parties at the 1997 British elections; the results were used by the British Broadcasting Corporation (BBC).

Public polls and surveys

The news media and several agencies conduct polls and surveys on the current topics of interest to the public. Presenting the results of a national poll, *The New York Times* (Tuesday, February 24, 1998) described how it was conducted. Interviews were obtained via telephone from 1153 adults in the U.S. Telephone exchanges were randomly selected from more than 42,000 residential exchanges across the country. To access both listed and unlisted numbers within each exchange, random digits were added to form complete telephone numbers. Within each selected household, one respondent was selected randomly to answer the questionnaire. The results were weighted to take account of household sizes and the number of telephone lines into the households and also to adjust for the variability in the sample geographical region, sex, race, age, and education.

In another survey conducted by *The New York Times*/CBS News Poll (*New York Times*, Wednesday, October 20, 1999), 1038 teenagers, 13 to 17 years old, were randomly selected from the 42,000 residential exchanges. The topics included their desire to get good grades, usage of computers and e-mail, need for having friends, problems with drugs and peer pressure, and concerns with violence and crime as well as with their safety.

Campus surveys

College campuses frequently conduct surveys on students regarding their educational programs, living quarters, dining facilities, sports activities, and similar topics of interest. Four of the surveys conducted during the sampling course at the University of Rochester on approximately 4000 undergraduate students are briefly described below.

As a project in the sampling course one year, a survey on the athletic facilities on the campus was conducted. All the students in the university were stratified according to the class—freshman, sophomore, junior, and senior, and the male–female classification. An overall sample of about 200 students, 5% of the total, was distributed to the eight groups proportional to their sizes. Samples were selected randomly from each of the eight groups, approximately 25 from each group, and information on the participation of the students in different types of sports and their visits to the gymnasium was obtained through telephone interviews. More than one attempt was needed to contact some of the sampled students on the telephone, and some of the contacted students either did not respond to the questions or provided only partial responses. Estimates for the numbers of students participating in different types of sports and other characteristics were made only after making suitable adjustments for the nonresponse. Another survey on the students included general topics of interest to the students related to studies, residence halls, meals, alcohol consumption, belief in God, and religious faith. In a food survey conducted on the campus, free samples of two or three items were available during the lunch hour, and the participants rated their preferences on a 1 to 5 scale.

In another food survey, students expressed their opinions on the quality of the food and the services at the three dining facilities. Inferences from these types of surveys are valid if the responses can be considered to be representative of all the students on the campus.

Exercises

1.1. In an assembly or gathering of the public, sometimes the loudness of the applause or the show of hands is used for assessing the approval of a proposition of general interest. When can such a procedure provide a valid estimate for the entire population?

1.2. Consumers shopping at supermarkets are frequently offered samples of food and beverages and their preferences are analyzed. Can the estimates from such analyses be generalized to the entire community?

1.3. People's choices for household purchases are sometimes noted by contacting the persons passing a central location in a shopping mall, for example, one every 5 minutes. Present two situations suitable for generalizing the results of the above type of information to the entire population.

1.4. The opinions of students on the dining facilities at a university were collected during 1 week at the four dining centers on the campus. A total of 300 students expressed their opinions; 90, 80, 65, and 65 of the freshman to senior classes, respectively. The 4000 students consisted of approximately equal numbers in each class. (a) Are the estimates from this type of survey valid for the 4000 students? (b) Can the responses be generalized separately to each of the four classes? (c) Instead of the above procedure, how would one plan to obtain the opinions from a sample of 300 students?

1.5. To estimate the average expenses of approximately 4000 college students for books, entertainment, and travel, the mean of a 5% random sample from all the students can be considered. Alternatively, the sample can be distributed in the four classes, freshmen to seniors, proportional to their sizes and their means weighted accordingly. Which of these two procedures can provide better estimates?

1.6.   The average weight of 30 children can be estimated by
       the mean weight of a random sample of six of the chil-
       dren. Alternatively, as they stand according to their
       heights, the mean weight of (a) the shortest and tallest
       child or (b) the two middle ones may be used for esti-
       mating the average weight. Which of these procedures can
       provide a valid estimate for the average weight of the 30
       children?

1.7.   The households in a region are numbered from 1 to 300.
       One can consider a 5% random sample from the house-
       holds to estimate, for example, characteristics related
       to their education, health, and income. Alternatively,
       1 household from the first 20 can be randomly selected
       and every 20th household from the selected household
       can be included in the sample. For which situations
       would each of these two procedures be recommended?

1.8.   As another alternative to the two procedures in Exercise 1.7,
       the households can be randomly divided into 20 groups with
       15 in each, and one of the groups can be chosen randomly.
       When is this procedure preferred?

1.9.   Purchases of several products such as refrigerators, air-
       conditioners, radios, televisions, bicycles, and sporting
       goods are accompanied by registration and warranty
       forms. They contain questions related to family size, ages
       of the household members, educational and income levels,
       and so forth. The completed and returned forms are some-
       times used to estimate the demographic and socioeconomic
       characteristics of the population and their preferences for
       the above types of products. Describe two situations for
       which these practices provide valid estimates.

1.10. In a *straw poll* of the 400 persons at an election meeting,
       200 were in favor of the first candidate, 160 the second
       candidate, but the remaining 40 did not express any
       opinion. If the 400 can be considered to be representative
       of all the registered voters, for which situations can the
       following procedures be recommended? (a) Ignore the 40
       without any opinion and estimate the preferences from
       the 360 responses, (b) add 20 to each of the candidates
       and consider 220 and 180 to be in favor of the two
       candidates, and (c) add (200/400)40 and (160/400)40 to
       the first and second candidates.

1.11.   To estimate the percentages of the different brands of computers purchased in a metropolitan area and the average price paid for each brand, two lists were available: (a) the list of all the purchases from the computer stores and (b) the lists of all the purchases from the computer stores, universities, general stores, and through the mail. The second list has twice as many purchases as the first. Information on the purchased brands and the prices paid were obtained from samples of sizes 100 and 200, respectively, from the two lists, and 30 of the purchases appeared in both the samples. From the information observed in these samples, how would one estimate the percentage purchases for the brands and the average of the prices paid.

1.12.   For the information collected through the two samples in Exercise 1.11, how would the difference in the average of the prices for the computer stores and other places be estimated?

CHAPTER 2

Simple Random Sampling: Estimation of Means and Totals

2.1 Introduction

Simple random sampling is the simplest probability sampling procedure. In this method of selecting a sample of the population units, every sample of a fixed size is given an equal chance to be selected. Every population unit is given an equal chance of appearing in the sample. Similarly, every pair of units has an equal chance of appearing in the sample. In general, every collection of units of a fixed size has an equal chance of being selected.

In the following sections, procedures for selecting a simple random sample and for using it to estimate the population means, totals, and variances of the characteristics of interest are presented. Properties of this sampling procedure are studied in detail. Methods for determining the sample size required for a survey are also examined.

2.2 Population total, mean, and variance

For the sake of illustration, the verbal and math scores in the Scholastic Aptitude Test (SAT) for a small population of $N = 6$ students are presented in Table 2.1. One can denote the **population units**, students in this illustration, by $U_1, U_2, ..., U_N$ or briefly as $U_i, i = 1, 2, ..., N$. The verbal scores of these units can be denoted by $x_i, i = 1, 2, ..., N$ and the math scores by $y_i, i = 1, 2, ..., N$.

Total and mean

For the y-characteristic, the population total and mean are given by

$$Y = \sum_1^N y_i = (y_1 + y_2 + \cdots + y_N) \tag{2.1}$$

Table 2.1. SAT verbal and math scores.

Student i	Verbal x_i	Math y_i
1	520	670
2	690	720
3	500	650
4	580	720
5	530	560
6	480	700
Total	3300	4020
Mean	550	670
Variance		
σ^2	4866.67	3066.67
S^2	5840	3680
S	76.42	60.66
C.V. (%)	13.89	9.05

C.V. = coefficient of variation.

and

$$\bar{Y} = Y/N. \tag{2.2}$$

The mean is a measure of the center or **location** of the N population units. From Table 2.1, for the math scores, $Y = (670 + 720 + \cdots + 720) = 4020$ and $\bar{Y} = 4020/6 = 670$.

Variance and standard deviation (S.D.)

The variance of the y-characteristic is defined as

$$\sigma^2 = \frac{\sum_1^N (y_i - \bar{Y})^2}{N} = \frac{(y_1 - \bar{Y})^2 + (y_2 - \bar{Y})^2 + \cdots + (y_N - \bar{Y})^2}{N}$$

$$= \frac{\sum_1^N y_i^2 - N\bar{Y}^2}{N}. \tag{2.3}$$

or as

$$S^2 = \frac{\sum_1^N (y_i - \bar{Y})^2}{N - 1}.$$ (2.4)

For convenience, in sample surveys, the expression in (2.4) is frequently used for the variance. Either of the above expressions can be obtained from the other, since $S^2 = (N - 1)\sigma^2/N$.

The standard deviation, σ or S, is obtained from the positive square roots of (2.3) and (2.4). The variance and standard deviation describe the dispersion or spread among the observations of the population units. For the math scores, $S^2 = [(670 - 670)^2 + (720 - 670)^2 + \cdots (700 - 670)^2]/5 = 3680$ and $S = 60.7$. An alternative expression for the variance is presented in Exercise 2.12, which explicitly expresses the variance as the average of the squared differences of all the pairs of observations.

Coefficient of variation (C.V.)

The C.V. of a characteristic is the ratio of the standard deviation to its mean. This index, unlike the standard deviation and the variance, is not affected by the unit of measurement. For example, it will be the same if the incomes of a group of people are measured in thousands of dollars or tens of thousands of dollars. Similarly, the C.V. will be the same if the incomes are expressed in dollars, British pounds, or any other currency. For the math scores, it is $(S/\bar{Y}) = (60.66/670) = 0.0934$, or 9.34%.

For the x-characteristic, the mean, variance, standard deviation, and C.V. are defined similarly. They are presented in Table 2.1 for both the math and verbal scores. Although the mean is larger for the math scores, the standard deviation and C.V. are larger for the verbal scores.

2.3   Sampling without replacement

Consider estimating the mean and variance of the math scores by selecting a sample of size $n = 2$ randomly from the $N = 6$ students. One can select the sample by writing the names, numbers, or labels of the six students on pieces of paper or cards, thoroughly mixing them, selecting one **randomly**, setting it aside, and then selecting the second one **randomly** from the remaining five. This is one of the procedures for selecting a sample randomly **without replacement**. Alternatively, one can list all the $_6C_2$ (six choose two), that is, $6!/2!(6 - 2)! = 15$ possible samples and select one of the samples randomly.

Random samples from a population, especially large, can also be selected from random number tables or computer software packages. A set of random numbers generated through a computer program is presented in Table T1 in the Appendix. In any table of random numbers, each of the digits from zero to 9 has a 1 in 10 chance of appearing. To select a sample of size 20 from a population of 200 units, for example, first label the units from 0 to 199. Start with any three columns together anywhere in the table, select some numbers, rejecting a number that exceeds 199 and that has already been selected, move to another set of three columns, select some more numbers in the same manner, and continue the procedure until 20 numbers are selected. The numbers can also be selected, three at a time, from the rows or both from the rows and columns of the random numbers. For example, a sample of two from the population of six units can be selected by starting with a single row or column in the random number table. Throughout the chapters, a sample selected randomly without replacement is referred to as a **simple random sample**.

As will be seen in the following sections, random sampling and statistical procedures enable one to estimate population quantities such as the total and mean, assess the error in the estimates, and also find intervals for these quantities for specified probabilities.

2.4   Sample mean and variance

The observations of a sample of size n can be denoted by y_i, $i = 1, 2, \ldots, n$. The **sample mean** and **variance** of the y-characteristic are

$$\bar{y} = \sum_1^n y_i/n = (y_1 + y_2 + \cdots + y_n)/n \tag{2.5}$$

and

$$s^2 = \frac{\sum_1^n (y_i - \bar{y})^2}{n - 1}. \tag{2.6}$$

By either of the procedures described in Section 2.3, if units (U_1, U_5) are selected, for the math scores $\bar{y} = (670 + 560)/2 = 615$ and $s^2 = [(670 - 615)^2 + (560 - 615)^2]/(2 - 1) = 6050$. These sample **estimates** 615 and 6050 differ from 670 and 3680, the actual population mean and variance. These differences should be anticipated, since the sample is only a fraction of the population. For increased sample sizes, the differences can be small.

2.5   Properties of simple random sampling

For this type of sampling, the probability of

1. U_i, $i = 1, 2,...,N$ appearing at any draw of the sample is $P_i = 1/N$,
2. U_i appearing at a specified draw conditional on U_j $(i \neq j)$ appearing at another draw is $1/(N - 1)$, and
3. U_i and U_j appearing at two specified draws is $P_{ij} = 1/N$ $(N - 1)$.

The last result follows from (1) and (2).

To illustrate further properties of this type of sampling, 15 samples along with their means, variances, and summary figures for the math scores are presented in Table 2.2. Note, however, that in practice only one of these samples is actually selected. Two additional properties of simple random sampling are as follows.

1.   There are $_NC_n$ possible samples and each population unit U_i, $i = 1, 2,...,N$, appears in $_{(N-1)}C_{(n-1)}$ samples. Hence, the probability of a population unit appearing in the samples is $_{(N-1)}C_{(n-1)}/_NC_n = n/N$. In this illustration, each of the six units appears in 5 of the 15 samples and hence it has a chance of 1/3 of appearing in the samples.
2.   Each pair of the N population units appears in $_{(N-2)}C_{(n-2)}$ of the samples and has a chance of $_{(N-2)}C_{(n-2)}/_NC_n = n(n - 1)/N$ $(N - 1)$ of appearing in the samples. In the illustration, each pair has a chance of 1/15 of appearing in the samples.

Similar results hold for the appearance of different combinations of the population units in the sample of any size selected through simple random sampling.

Table 2.2. All possible samples of size two; means and variances of the math scores.

Sample	Units	Observations	Mean	Variances	S.D.
1	1, 2	670, 720	695	1,250	35.36
2	1, 3	670, 650	660	200	14.14
3	1, 4	670, 720	695	1,250	35.36
4	1, 5	670, 560	615	6,050	77.78
5	1, 6	670, 700	685	450	21.21
6	2, 3	720, 650	685	2,450	49.50
7	2, 4	720, 720	720	0	0
8	2, 5	720, 560	640	12,800	113.14
9	2, 6	720, 700	710	200	14.14
10	3, 4	650, 720	685	2,450	49.50
11	3, 5	650, 560	605	4,050	63.64
12	3, 6	650, 700	675	1,250	35.36
13	4, 5	720, 560	640	12,800	113.14
14	4, 6	720, 700	710	200	14.14
15	5, 6	560, 700	630	9,800	99.00
Expected value of the sample mean			670		
Expected value of the sample variance				3,680	
Expected value of the sample standard deviation					49.03

2.6   Unbiasedness of the sample mean and variance

Sample mean

The mean of each of the possible simple random samples can be denoted by $\bar{y}_k$, $k = 1,...,_NC_n$, with the associated probability $P_k = 1/_NC_n$ for each. Following the definition in Appendix A2.1, the expected value of the sample mean is

$$E(\bar{y}) = \sum_k P_k \bar{y}_k = \frac{1}{\binom{N}{n}} \frac{1}{n} \left[\left(\sum_1^n y_i \right)_1 + \left(\sum_1^n y_i \right)_2 + \cdots + \left(\sum_1^n y_i \right)_t \right], \quad (2.7)$$

where $t = {}_NC_n$.

Since every population unit appears in $_{(N-1)}C_{(n-1)}$ of the possible $_NC_n$ samples, the summation in the square brackets is equal to

$_{(N-1)}C_{(n-1)}(\Sigma_1^N y_i)$. With this expression,

$$E(\bar{y}) = \bar{Y}. \tag{2.8}$$

Thus, the sample mean $\bar{y}$ is **unbiased** for the population mean $\bar{Y}$. As a verification, the average of the 15 sample means of the math scores in Table 2.2 is identically equal to the population mean.

Sample variance

The sample variance in (2.6) can be expressed as

$$s^2 = \frac{1}{n-1}\left[\sum_1^n y_i^2 - n\bar{y}^2\right]. \tag{2.9}$$

Appendix A2.3 shows that

$$E(s^2) = \frac{1}{N-1}\left[\sum_1^N y_i^2 - N\bar{Y}^2\right] = S^2. \tag{2.10}$$

Thus, s^2 is unbiased for S^2. For the math scores, as can be seen from Table 2.2, the average of the 15 sample variances is equal to the population variance.

Note that if all N population values are known, there is no need for sampling to estimate the total, mean, variance, or any other population quantity. Second, in practice only one sample of size n is selected from the population. All six population values are presented in Table 2.1 and all the possible samples of size two are presented in Table 2.2 just to illustrate and demonstrate the important properties of simple random sampling. The unbiasedness of the sample mean and variance as seen through (2.8) and (2.10) are theoretical results, and they are verified through Table 2.2. The samples in this table are used to illustrate other properties of this sampling procedure.

2.7   Standard error of the sample mean

A large portion of the 15 sample means in Table 2.2 differ from each other, although their expected value is the same as the population mean.

The variance among all the $_NC_n$ possible sample means is

$$V(\bar{y}) = E(\bar{y} - \bar{Y})^2 = E(\bar{y}^2) - 2\bar{Y}E(\bar{y}) + \bar{Y}^2 = E(\bar{y}^2) - \bar{Y}^2. \quad (2.11)$$

As shown in Appendix A2.4, this variance can be expressed as

$$V(\bar{y}) = \frac{N-n}{Nn}S^2 = (1-f)\frac{S^2}{n}, \quad (2.12)$$

where $f = n/N$ is the **sampling fraction** and $(1-f)$ is the **finite population correction** (fpc).

The standard deviation of $\bar{y}$, known as its **standard error** (S.E.), is obtained from

$$\text{S.E.}(\bar{y}) = \sqrt{V(\bar{y})} = \sqrt{\frac{(1-f)}{n}}\,S. \quad (2.13)$$

For the math scores, since $S^2 = 3680$, the variance of $\bar{y}$ is equal to $(6-2)(3680)/12 = 1226.67$, and hence it has an S.E. of 35.02.

As can be seen from the definition in (2.11) and (2.13), $V(\bar{y})$ and S.E.$(\bar{y})$ measure the average departure of the sample mean from the population mean. Further, as the S.E.$(\bar{y})$ becomes small, $\bar{y}$ becomes closer to $\bar{Y}$. This observation can also be made from replacing the random variable X by $\bar{y}$ in the Tschebycheff inequality in Appendix A2.2. Notice from (2.13) that the S.E.$(\bar{y})$ will be small if the sample size is large. If the population size is large, the variance in (2.12) and the S.E. in (2.13) become S^2/n and $S/\sqrt{n}$, which will not differ much from σ^2/n and $\sigma^2/\sqrt{n}$.

Since s^2 is unbiased for S^2, an unbiased estimator for $V(\bar{y})$ is obtained from

$$v(\bar{y}) = \frac{(1-f)}{n}s^2. \quad (2.14)$$

From the sample observations, the S.E. in (2.13) can be estimated from $(1-f)s/\sqrt{n}$.

With the sample units (U_1, U_5), for the math scores, $s^2 = 6050$. From (2.14), $v(\bar{y}) = [(6-2)/6 \times 2](6050) = 2016.67$, and hence the S.E. of $\bar{y}$ from the sample is equal to 44.91.

Notice from Table 2.2 that the variances of the 15 possible samples vary widely, although their expected value is the same as the population variance. Thus, the sampling error of s^2 for estimating S^2 can be large, unless the sample size is large.

2.8 Distribution of the sample mean

The distribution of the 15 sample means of the math scores in Table 2.2 is presented in Table 2.3. Grouping of the means as in Table 2.4 into six classes of equal width of 20 results in an approximation to this distribution. The histogram of the distribution with the mid-values 602.5, 627.5, 652.5, 677.5, 702.5, and 727.5 and the frequencies is presented in Figure 2.1.

For an infinite population, from the **central limit theorem**, the sample mean follows the normal distribution provided the sample size is large. As a result, $Z = (\bar{y} - \bar{Y})/\text{S.E.}(\bar{y})$ approximately follows the standard normal distribution with mean zero and variance unity. For this distribution, which is tabulated and also available through computer software programs, $(1 - \alpha)\%$ of the probability (area) lies between $-Z_{\alpha/2}$ and $Z_{\alpha/2}$. For example, 90% of the area lies between -1.65 and 1.65, 95% between -1.96 and 1.96 or roughly between -2 and 2, and 99% between -2.58 and 2.58.

Table 2.3. Distribution of the Sample mean.

Mean	Frequency	Mean	Frequency
605	1	675	1
615	1	685	3
630	1	695	2
640	2	710	2
660	1	720	1

Table 2.4. Frequency distribution.

Mean	Frequency
590–615	1
615–640	2
640–665	3
665–690	4
690–715	4
715–740	1

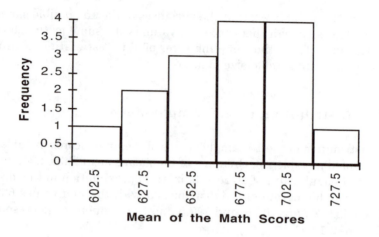

Figure 2.1. Approximate distribution of the sample mean for the math scores.

2.9   Confidence limits for the mean

The Tschebycheff inequality in Appendix A2.2 may be used for this purpose, by replacing X by the sample mean $\bar{y}$. With $\gamma = 2$ for example, 75% of the sample means falls within $2\text{S.E.}(\bar{y})$ of the population mean $\bar{Y}$.

With the normal approximation, the probability of $(\bar{y} - \bar{Y})/\text{S.E.}(\bar{y})$ falling between $-Z_{\alpha/2}$ and $Z_{\alpha/2}$ is $(1 - \alpha)$, that is,

$$P[\bar{y} - Z_{\alpha/2}\text{S.E.}(\bar{y}) < \bar{Y} < \bar{y} + Z_{\alpha/2}\text{S.E.}(\bar{y})] = (1 - \alpha). \qquad (2.15)$$

From this expression, the **lower** and **upper confidence limits** for $\bar{Y}$ are obtained from $C_L = \bar{y} - Z\,\text{S.E.}(\bar{y})$ and $C_U = \bar{y} + Z\,\text{S.E.}(\bar{y})$, and the **confidence width** is given by $(C_U - C_L) = 2Z\,\text{S.E.}(\bar{y})$; the subscript $\alpha/2$ is suppressed for convenience. Confidence limits with small widths are desirable; otherwise, they provide only vague information regarding the unknown population mean $\bar{Y}$. This width is small if $\text{S.E.}(\bar{y})$ is small, that is, if the sample size is large.

For the math scores, if (U_1, U_5) are selected into the sample, 95% confidence limits for the population mean are $C_L = 615 - 1.96(35.02) = 546.36$, $C_U = 615 + 1.96(35.02) = 683.64$, and the confidence width is equal to 137.28.

The above limits are obtained with the known S^2. When this variance is estimated from the sample, as seen earlier, S.E.$(\bar{y})$ = 44.91. With this estimate, 95% confidence limits for the population mean are $C_L = 615 - 1.96(44.91) = 526.98$ and $C_U = 615 - 1.96(44.91) = 703.02$. The confidence width now is equal to 176.04.

For a large population, assuming that the characteristic y follows a normal distribution with mean $\bar{Y}$ and variance S^2, estimating S.E.$(\bar{y})$ from the sample, $t_{n-1} = (\bar{y} - \bar{Y})/$S.E. $(\bar{y})$ follows Student's t-distribution with $(n - 1)$ degrees of freedom (d.f.). The percentiles of this distribution are tabulated and are also available in software programs for statistical analysis. Now, the $(1 - \alpha)$% confidence limits for $\bar{Y}$ are obtained from $\bar{y} - t$ S.E.$(\bar{y})$ and $\bar{y} + t$ S.E.$(\bar{y})$, with the value of t corresponding to the $(1 - \alpha)$ percentile of the t-distribution with $(n - 1)$ d.f. If the sample size is also large, this percentage point does not differ much from that of the normal distribution, for example, 1.96 when $\alpha = 0.05$.

2.10 Estimating the total

It is frequently of interest to estimate the total Y of characteristics such as the household incomes and expenditures in a region, production of a group of industries, sales of computers in a city or state, and so on. An estimator of Y is

$$\hat{Y} = N\bar{y} = \frac{N}{n} \sum_{1}^{n} y_i. \qquad (2.16)$$

Since $E(\bar{y}) = \bar{Y}$, $E(\hat{Y}) = N\bar{Y} = Y$ and hence $\hat{Y}$ is unbiased for Y. Note that $\hat{Y}$ is obtained by multiplying, or **inflating**, the sample total $\sum_{1}^{n} y_i$ by (N/n), which is the reciprocal of the probability of selecting a unit into the sample.

The variance of $\hat{Y}$ is

$$V(\hat{Y}) = N^2 V(\bar{y}) = \frac{N(N - n)}{n} S^2. \qquad (2.17)$$

An unbiased estimator of this variance is given by

$$v(\hat{Y}) = N^2 v(\bar{y}) = \frac{N(N - n)}{n} s^2. \qquad (2.18)$$

The standard error of $\hat{Y}$ is obtained from

$$\text{S.E.}(\hat{Y}) = \sqrt{\frac{N(N-n)}{n}} S, \qquad (2.19)$$

and it is estimated by replacing S with the sample standard deviation s.

For the math scores, with the sample units (U_1, U_5), $\hat{Y} = 6(615) = 3690$. With the known value of S^2, $V(\hat{Y}) = 36(1226.67) = 44160.12$ and S.E.$(\hat{Y}) = 210.14$. When S^2 is estimated by s^2, $v(\hat{Y}) = 36(2016.67) = 72600.12$ and S.E.$(\hat{Y}) = 269.44$.

As in the case of the mean, $(1 - \alpha)\%$ confidence limits for Y are obtained from

$$\hat{Y} - Z \text{ S.E.}(\hat{Y}), \hat{Y} + Z \text{ S.E.}(\hat{Y}). \qquad (2.20)$$

These limits can also be obtained by multiplying the limits for the mean by N. If the observations follow the normal distribution, as before, Z is replaced by the percentile of the t-distribution.

For the math scores, when S^2 is known, with the sample units (U_1, U_5), 95% confidence limits for the total are $3690 - 1.96 (210.14) = 3278.13$ and $3690 + 1.96 (210.14) = 4101.87$, with the width of 823.74. With the estimated variance, these limits become $3690 - 1.96 (269.44) = 3161.90$ and $3690 + 1.96 (269.44) = 4218.10$, and now the confidence width is equal to 1056.20. Except for the round-off errors, the same limits and widths are obtained by multiplying the limits and widths for the population mean obtained in the last section by the population size, six in this illustration.

2.11   Coefficient of variation of the sample mean

The C.V. of the sample mean $\bar{y}$ is

$$\text{C.V.}(\bar{y}) = \text{S.E.}(\bar{y})/\bar{Y} = \sqrt{\frac{(1-f)}{n}}(S/\bar{Y}), \qquad (2.21)$$

which expresses the S.E. of $\bar{y}$ as a percentage of its expected value, the population mean $\bar{Y}$. To estimate this C.V., $(S/\bar{Y})$ is replaced by $(s/\bar{y})$. Note that for large n, the C.V. of $\bar{y}$ is obtained by dividing the C.V. of y by n. Second, the C.V. of $\hat{Y}$, $\sqrt{V(\hat{Y})}/Y$, is the same as that of $\bar{y}$.

As the variance of $\bar{y}$ and its S.E., the C.V. in (2.21) is also a measure of the **sampling error** of the mean. For the math scores, C.V. of $\bar{y}$ is $(35.02/670) = 0.0523$ or 5.23% when S^2 is known, and it is equal to $(44.91/615) = 0.0730$ or 7.30% with the estimated variance.

2.12   Sample size determination

The sample size for every survey has to be determined before it goes into operation. To estimate $\bar{Y}$ or Y, the size of the sample is usually needed for a prescribed value of the (1) S.E. of $\bar{y}$, (2) error of estimation, (3) confidence width, (4) C.V. of $\bar{y}$, or (5) the relative error, which is given by $|\bar{y} - \bar{Y}|/\bar{Y}$. These types of prescriptions are frequently employed in industrial research and commercial surveys. Information regarding the standard deviation S should be known for these critera. Past data or a preliminary sample of small size, also known as the pilot survey, can be used for this purpose.

Precision

The precision of an estimator is measured by the reciprocal of its variance. Any requirements for this precision can be translated into a prescription on the variance, and vice versa.

If the variance of $\bar{y}$ should not exceed a given value V, the sample size is obtained from

$$V(\bar{y}) \le V. \tag{2.22}$$

If N is large, this inequality can be expressed as $S^2/n \le V$, and with equality the required sample size is obtained from

$$n_l = \frac{S^2}{V}. \tag{2.23}$$

If N is not large, from (2.22), the smallest value of the required sample size is given by

$$n = \frac{n_l}{1 + (n_l/N)} \tag{2.24}$$

As expected, larger sample sizes are needed to estimate the mean with smaller S.E.

Error of estimation

With prior knowledge about $\bar{Y}$, it can be required that $\bar{y}$ should not differ from it by more than a specified amount of absolute error e, a small quantity. This requirement, however, can be satisfied only with a chance of $(1 - \alpha)$. Formally, this requirement can be expressed as the probability statement,

$$P_r\{|\bar{y} - \bar{Y}| \le e\} = (1 - \alpha). \tag{2.25}$$

From the correspondence of this probability with the Tschebycheff inequality in Appendix A2.2, $\gamma^2 = (1/\alpha)$ and $\gamma^2 V(\bar{y}) = e^2$. Solving these equations, n is obtained from (2.24) with $n_l = (\gamma S/e)^2$.

Alternatively, note that the above probability statement is the same as

$$P_r[|\bar{y} - \bar{Y}|/\text{S.E.}(\bar{y}) \le e/\text{S.E.}(\bar{y})] = (1 - \alpha). \tag{2.26}$$

With the assumption of normality, as noted before, the term on the left side in the brackets follows the standard normal distribution. As a result, $Z = e/\text{S.E.}(\bar{y})$ or $Z^2 V(\bar{y}) = e^2$, where Z is the $(1 - \alpha)$ percentile of this distribution. Now, n is obtained from (2.24) with

$$n_l = (ZS/e)^2. \tag{2.27}$$

As can be expected, the sample size will be large if the specified error e is small.

Confidence width

If it is required that the confidence width for $\bar{Y}$, with a confidence probability of $(1 - \alpha)$, should not exceed a prescribed amount w, the sample size is obtained from

$$2Z\,\text{S.E.}(\bar{y}) \le w. \tag{2.28}$$

Solution of this equation results in (2.24) for the sample size with

$$n_l = 4(ZS/w)^2. \qquad (2.29)$$

Coefficient of variation

If it is desired that the C.V. of $\bar{y}$ should not exceed a given value C_0, n should be obtained from

$$[(1 - f)/n]C^2 \le C_0^2, \qquad (2.30)$$

where $C = S/\bar{Y}$ is the population C.V.
  The required sample size is again obtained from (2.24) with

$$n_l = (C/C_0)^2. \qquad (2.31)$$

Note that this criterion can be satisfied only with some knowledge of the population C.V. Larger sample sizes are needed to estimate the population mean with a smaller C.V.

Relative error

The error of estimation relative to the population mean is $|\bar{y} - \bar{Y}|/\bar{Y}$. The requirement that this error should not exceed a prescribed value **a**, except for a chance of $(1 - \alpha)$, can be expressed as

$$\Pr\{|\bar{y} - \bar{Y}|/\text{S.E.}(\bar{y}) \le a\bar{Y}/\text{S.E.}(\bar{y})\} = (1 - \alpha) \qquad (2.32)$$

with the normality assumption, $Z^2 = a^2\bar{Y}^2/V(\bar{y})$. Solving this equation, n is obtained from (2.24) with

$$n_l = (ZC/a)^2. \qquad (2.33)$$

The population C.V., $C = S/\bar{Y}$, should be known for this criteria also.

  Example 2.1. Sample size: Consider a camera club with 1800 members, where it is required to estimate the average number of rolls of film used during a year. Consider also the information from the past that the average and standard deviation of the number of rolls of film have been around 6 and 4, respectively.

a.   To estimate $\bar{\bar{Y}}$ with an S.E. not exceeding 0.5, $n_l = 16/(0.5)^2 = 64$
from (2.23) and $n = 62$ from (2.24).
b.   To estimate $\bar{Y}$ with an error not exceeding 1, except for a 5% chance,
with the normal approximation, $n_l = (1.96)^2 16/1 = 62$ from (2.27)
and $n = 60$ from (2.24).
c.   With $w = 2$ and $(1 - \alpha) = 0.95$, $n_l = 62$ from (2.29) and $n = 61$
from (2.24).
d.   With the information on the mean and standard deviation, $C = 4/6$ or 67% approximately. If the C.V. of $\bar{y}$ should not exceed 8%,
$n_l = (0.67/0.08)^2 = 70$ from (2.31) and $n = 68$ from (2.24).
e.   If the relative error should not exceed (1/10) except for a 5% chance,
$n_l = (1.96)^2(0.67)^2/0.01 = 173$ from (2.33) and $n = 158$ from (2.24).

2.13   Sample sizes for individual groups

Frequently, it becomes important to estimate the means and totals for
two or more groups or subpopulations of a population. The sample
sizes for the groups may be found through specifications of the type
described in Section 2.12. Alternatively, the S.E., errors of estimation,
and confidence widths for each of the groups for suitable division of
the overall sample size required for the entire population may be
examined. These two procedures are illustrated in the following example. Chapters 5 and 6 examine in detail sample size determination for
two or more groups and also their differences.

Example 2.2. Sample sizes for groups: Consider the 1800 members in
Example 2.1 to consist of 1200 old and 600 new members with the
standard deviations of 2 and 4 for the number of cartons of film. As in
Example 2.1(b), sample sizes required to estimate the means of these
groups with the error of estimation not exceeding 1 for each group can
be found. Now, for the old group, $n_l = (1.96)^2 \times 4/1 = 15.4$ and $n = 15$
approximately. Similarly, for the new group, $n_l = (1.96)^2 \times 16/1 = 61.5$ and
$n = 56$ approximately. The total sample size of 71 required now is larger
than the size of 62 for the entire population, found in Example 2.1(b).

Exercises

2.1.   Among the 100 computer corporations in a region, average
of the employee sizes for the largest 10 and smallest 10
corporations were known to be 300 and 100, respectively.
For a sample of 20 from the remaining 80 corporations, the
mean and standard deviation were 250 and 110, respectively.

For the total employee size of the 80 corporations, find the (a) estimate, (b) the S.E. of the estimate, and (c) the 95% confidence limits.

2.2. Continuing with Exercise 2.1, for the average and total of the 100 corporations, find the (a) estimate, (b) the S.E. of the estimate, and (c) the 95% confidence limits.

2.3. During the peak season, the mean and standard deviation of the late arrivals for a sample of 30 flights of one airline were 35 and 15 minutes, respectively. For the average delay of all the flights of this airline, find (a) the 95% confidence limits with the t-distribution and (b) the C.V. of the sample mean.

2.4. As in Exercise 2.3, the mean and standard deviation of the late arrivals for a sample of 30 flights of another airline were 30 and 20, respectively. Which of these airlines would one prefer if (a) the upper 95% confidence limit for the average with the t-distribution should not exceed 40 minutes and (b) the coefficient of variation for the sample mean should not exceed 10%?

2.5. The total profit for $N_1 = 5$ of the largest computer companies was known to be $Y_1 = 500$ (million dollars). For a sample of $n_2 = 9$ from the remaining $N_2 = 45$ computer companies, the mean and standard deviation were $\bar{y}_2 = 30$ and $s_2 = 15$. An estimate suggested for the total profit of the $N = 50$ companies was $t_1 = N(Y_1 + n_2\bar{y}_2)/(N_1 + n_2)$. From the above results, $t_1 = (50/14)(500 + 9 \times 30) = 1882.14$. (a) Find an expression for the variance of this estimator, and find its S.E. from the sample observations. (b) Is this an unbiased estimator for the total profit?

2.6. Another estimator for the total profit of the 50 companies in Exercise 2.5 is $t_2 = Y_1 + N_2\bar{y}_2$, and from the sample observations it is equal to $500 + 45 \times 30 = 1850$. (a) Find an expression for the variance of this estimator and find its S.E. from the sample. (b) Is this an unbiased estimator for the total profit?

2.7. Find the sample size required for the problem in Example 2.1 if (a) the S.E. should not exceed 0.25, (b) the sample mean should not differ from the actual mean by more than one roll of film, except for 10% chance, and (c) the relative error should not exceed 7% except for a 5% chance.

2.8. The metropolian area and the suburbs together in a region consist of 5, 10, and 10 thousand families with one, two, and three or more children. For these three types of families, preliminary estimates of the averages and standard deviations of the number of hours of television watching in a week are (10, 15, 20) and (6, 10, 15), respectively. Find the sample sizes required to estimate the above average for each group if the error of estimation should not exceed 2 hours in each case, except for a 5% chance.

2.9. Continuing with Exercise 2.8, find the sample sizes if the error of estimation should not exceed 10% of the actual average in each case, except for a 5% chance.

2.10. *Project.* Select all the 20 samples of size three from the population of six students in Table 2.1 without replacement and verify the five properties described in Section 2.5 for the probabilities of selecting the units and their appearance in the sample. From each sample, find the 95% confidence limits for the population mean of the math scores with the known population variance and its estimates; use the normal deviate $Z = 1.96$ in both cases. (a) For both the procedures, find the proportion of the confidence intervals enclosing the actual population mean, that is, the **coverage probability**. (b) Compare the average of the confidence widths obtained with the estimates of variance with the exact width for the case of known variance.

2.11. (a) Show that $\Sigma_{i<j}^{N}\Sigma^{N}(y_i - y_j)^2 = N\Sigma_1^{N}(y_i - \bar{Y})^2$, and hence S^2 in (2.4) is the same as $\Sigma_{i<j}^{N}\Sigma^{N}(y_i - y_j)^2/N(N - 1)$.
(b) Similarly, show that s^2 in (2.6) can be expressed as $\Sigma_{i<j}^{N}\Sigma^{N}(y_i - y_j)^2/n(n - 1)$.
(c) From these expressions, show that for simple random sampling without replacement, s^2 is unbiased for S^2.

2.12. For simple random sampling without replacement, starting with the expectation of $\Sigma_1^{n}(y_i - \bar{Y})^2$, show that $V(\bar{y}) = (1 - f)S^2/n$.

Appendix A2

Expected value, variance, and standard deviation

The expected value of a random variable X taking values $(x_1, x_2,...,x_k)$ with probabilities $(p_1, p_2,...,p_k)$, $p_1 + p_2 + \cdots + p_k = 1$, is

$$\mu = E(X) = x_1 p_1 + x_2 p_2 + \cdots + x_k p_k.$$

For two constants c and d, the expected value of $Y = (cX + d)$ is

$$E(cX + d) = cE(X) + d = c\mu + d.$$

The variance of X is

$$\sigma^2 = V(X) = E(X - \mu)^2 = (x_1 - \mu)^2 p_1 + (x_2 - \mu)^2 p_2 + \cdots$$
$$+ (x_2 - \mu)^2 p_k.$$

The variance can also be expressed as $E(X^2) - \mu^2$.

The standard deviation σ is obtained from the positive square root of the variance. The variance of $Y = cX + d$ is

$$V(Y) = E[y - E(Y)]^2 = E[(cX + d) - (c\mu + d)]^2$$
$$= c^2 E(X - \mu)^2 = c^2 V(x).$$

Adding a constant to a random variable does not change its variance. When the random variable is multiplied by a constant c, the variance becomes c^2 times the original variance.

Tschebycheff inequality

For a random variable X, this inequality is given by

$$\Pr[|X - E(X)| \leq \gamma\sqrt{V(X)}] \geq 1 - \frac{1}{\gamma^2},$$

where γ is a chosen positive quantity. For instance, if $\gamma = 2$, 75% of the observations fall within two standard deviations of $E(X)$, the population mean.

Unbiasedness of the sample variance

The sample variance in (2.9) can be expressed as

$$s^2 = \frac{1}{n-1}\left[\sum_1^n y_i^2 - n\bar{y}^2\right] = \frac{1}{n}\sum_1^n y_i^2 - \frac{1}{n(n-1)}\sum_{i \neq j}^n \sum^n y_i y_j.$$

Noting that each of the samples is selected with a probability of $1/_NC_n$, following the definition for the expectation,

$$E(s^2) = \frac{1}{\binom{N}{n}}\frac{1}{n}\left[\left(\sum_i y_i^2\right)_1 + \left(\sum_i y_i^2\right)_2 + \cdots + \left(\sum_i y_i^2\right)_t\right]$$

$$-\frac{1}{\binom{N}{n}}\frac{1}{n(n-1)}\left[\left(\sum_{i \neq j}\sum y_i y_j\right)_1 + \left(\sum_{i \neq j}\sum y_i y_j\right)_2 + \cdots + \left(\sum_{i \neq j}\sum y_i y_j\right)_t\right].$$

As in the case of the mean, the first summation is equal to $_{(N-1)}C_{(n-1)}$ $\sum_1^N y_i^2$. Similarly, the second summation is equal to $_{(N-2)}C_{(n-2)} \sum_{i \neq j}^N \sum^N y_i y_j$. Substituting these expressions,

$$E(s^2) = \frac{1}{N}\sum_1^N y_i^2 - \frac{1}{N(N-1)}\sum_{i \neq j}^N \sum^N y_i y_j$$

$$= \frac{1}{N-1}\left[\sum_1^N y_i^2 - N\bar{Y}^2\right] = S^2.$$

Variance of the sample mean

Since

$$\bar{y}^2 = \frac{1}{n^2}\left[\sum_i^n y_i^2 + \sum_{i \neq j}^n \sum^n y_i y_j\right],$$

$$E(\bar{y}^2) = \frac{1}{n^2}\left[\frac{n}{N}\sum_i^N y_i^2 + \frac{n(n-1)}{N(N-1)}\sum_{i \neq j}^N \sum^N y_i y_j\right]$$

$$= \frac{(N-n)}{nN(N-1)}\sum^N y_i^2 + \frac{(n-1)}{n(N-1)}\bar{Y}^2.$$

Now, with simplification,

$$V(\bar{y}) = E(\bar{y}^2) - \bar{Y}^2 = (1-f)S^2/n.$$

CHAPTER 3

Simple Random Sampling: Related Topics

3.1   Introduction

This chapter contains topics of general interest both in sample surveys and for any type of statistical analysis. The results presented in the appendix to this chapter are also of a general type, and they are widely applicable.

The last chapter demonstrated that the sample mean and variance are unbiased estimators for the population mean and variance. Several estimators in sample surveys take the form of the ratio of two sample means or they may be nonlinear in the observations. The property of unbiasedness may not hold for these types of estimators. In such cases, as described in the following section, in addition to the **variance** of an estimator, one can examine its **bias** and **mean square error** (MSE) for assessing its departure from the population quantity being estimated. The relevance of these concepts to the criteria of **precision**, **accuracy**, and **consistency** is described in Section 3.3.

Covariance and correlation between two characteristics and their sample means are examined in Section 3.4. In the next section, this topic is followed by the estimation of the mean and variance of a linear combination of two characteristics, the sum and difference in particular.

In several types of surveys, for convenience or otherwise, the sample is selected **systematically** from the population. Section 3.6 examines its similarity and difference from simple random sampling. In surveys on human populations, responses usually cannot be obtained from all the sampled units. The bias arising from **nonresponse** is briefly examined in Section 3.7. It is examined further in the exercises of some of the coming chapters, and Chapter 11 is entirely devoted to the various procedures available for counteracting the effects of nonresponse.

Chapter 2 noted that the procedures of confidence limits developed for infinite populations can be used as approximate procedures for finite populations. Section 3.8 considers the adoption of the classical **tests of hypotheses** for inferences regarding the means and totals of finite populations.

3.2   Bias, variance, and mean square error of an estimator

As an illustration, consider the sample standard deviation s to estimate the population standard deviation S. Although s^2 is *unbiased* for S^2, s is biased for S. The bias, variance, and MSE of the sample standard deviation s are

$$B(s) = E(s) - S, \tag{3.1}$$

$$V(s) = E[s - E(s)]^2 \tag{3.2}$$

and

$$\text{MSE}(s) = E(s - S)^2 = E[s - E(s) + E(s) - S]^2$$

$$= E[s - E(s)]^2 + [E(s) - S]^2$$

$$= V(s) + B^2(s). \tag{3.3}$$

Note that the expectation of the cross-product term in the first expression of (3.3) vanishes.

The bias of an estimator in general can be zero, positive, or negative. In the first case, it is unbiased for the population quantity being estimated. It overestimates in the second case and underestimates in the last.

For the math scores in Table 2.1, the population standard deviation is $S = 60.67$. The expected value of the sample standard deviation obtained from the average of the last column of Table 2.2 is $E(s) = (35.36 + 14.14 + \cdots + 99.00)/15 = 49.03$. Thus the bias of s is $B(s) = 49.03 - 60.7 = -11.64$. This is an illustration of the general result that the sample standard deviation underestimates the population standard deviation.

In some instances, the bias becomes negligible for large sample sizes. For example, consider $\hat{S}^2 = \Sigma_1^n(y_i - \bar{y})^2/n = (n-1)s^2/n$ as an alternative estimator for S^2. Since $E(\hat{S}^2) = (n - 1)S^2/n$, the bias of $\hat{S}^2$ is equal to $-S^2/n$, which becomes small as n increases. Suitable adjustments can be made to reduce or eliminate the bias in some estimators. In this case, note that $n\hat{S}^2/(n - 1)$ is the same as s^2, which is unbiased.

Selection of a sample haphazardly is frequently characterized as biased sampling, and faulty measurements are described as biased observations. In statistics, bias refers to the estimation of a population quantity with its formal definition as in (3.1).

As defined in (3.2) and (3.3), the variance and MSE, respectively, are the averages of the squared deviation of an estimator from its **expected value** and the **actual value**. For the math scores, from Table 2.2, $V(s) = (35.36 - 49.03)^2 + (14.14 - 49.03)^2 + \cdots + (99 - 49.03)^2 = 1276.44$ and $MSE(s) = (35.36 - 60.67)^2 + (14.14 - 60.67)^2 + \cdots + (99 - 60.67)^2 = 1411.79$. Except for the rounding errors, this value for the MSE can also be obtained by adding the squared bias to the variance. Note also that S.E.$(s) = (1276.44)^{1/2} = 35.73$, and the square root of the MSE is $(1411.79)^{1/2} = 37.57$.

One can examine the severity of the bias of an estimator by comparing its absolute value with the actual population quantity being estimated. For the sample standard deviation of the math scores, this relative value is 11.63/60.67, which exceeds 17%. The absolute bias of an estimator can also be compared with its S.E. or the square root of its MSE. For the math scores, these relative values, respectively, are $11.73/35.73 = 0.33$ or 33% and $11.73/37.57 = 0.31$ or 31%, both of which are very high.

Since the actual population quantity of such S is not known, one cannot examine the bias as above. As will be seen in some of the coming chapters, it is possible to examine the bias from its expression. Procedures for reducing or completely eliminating the bias of an estimator are presented in Chapter 12.

3.3   Precision, accuracy, and consistency

The **precision** of an estimator is inversely proportional to its variance. As can be seen from Equation 2.12, the precision of the sample mean increases with the sample size.

The **accuracy** of an estimator is inversely proportional to its MSE. It is desirable that any estimator is highly accurate. As can be seen from Equation 3.3, estimators that are highly precise but have a large positive or negative bias cannot have a high accuracy and will not have much practical utility.

For infinite populations, an estimator is defined to be **consistent** if it approaches the population quantity being estimated as the sample size becomes large. From the Tschebycheff inequality in A2.2, note that an unbiased estimator is consistent if its variance becomes small. In the case of a finite population, as noted in Section 2.7, the sample mean, which is unbiased, approaches the population mean if the sample size becomes close to the population size.

3.4   Covariance and correlation

In almost every survey, information is collected on more than one characteristic of the population. Incomes of couples, test scores of students in two or more subjects, and employee sizes, sales, and profits of corporations at several periods of time are some of the examples. Let (x_i, y_i), $i = (1,...,N)$, represent two of these characteristics. One can compute the covariance and correlation of these characteristics from the ungrouped or grouped data.

Ungrouped data

Similar to the total and mean of y_i in (2.1) and (2.2), let X and $\bar{X}$ denote the population total and mean of x_i. With the definition in (2.4), let S_x^2 and S_y^2 denote the variances of x_i and y_i. The population covariance and correlation coefficient of these characteristics are defined as

$$S_{xy} = \frac{\sum_1^N (x_i - \bar{X})(y_i - \bar{Y})}{N-1} = \frac{\sum_1^N x_i y_i - N\bar{X}\bar{Y}}{N-1} \tag{3.4}$$

and

$$\rho = \frac{S_{xy}}{S_x S_y}. \tag{3.5}$$

This coefficient ranges from -1 to 1. For high positive and negative correlation between x and y, it will be close to $+1$ and -1, respectively, and for low correlation it will be close to 0. Notice that the covariance can also be expressed as $S_{xy} = \rho\, S_x S_y$.

For the verbal and math scores in Table 2.1, $S_{xy} = [(520 - 550)(670 - 670) + (690 - 550)(720 - 670) + \cdots + (480 - 550)(700 - 670)]/5 = 1920$.

Since $S_x = 76.42$ and $S_y = 60.66$, $\rho = 1920/(76.42 \times 60.66) = 0.41$. Thus, there is a positive correlation between the math and verbal scores, but it is not very high.

For a simple random sample of size n, let $(\bar{x}, \bar{y})$ and (s_x^2, s_y^2) denote the means and variances of (x_i, y_i). The sample covariance and correlation coefficient are

$$s_{xy} = \frac{\sum_1^n (x_i - \bar{x})(y_i - \bar{y})}{n - 1} = \frac{\sum_1^n x_i y_i - n\bar{x}\bar{y}}{n - 1} \qquad (3.6)$$

and

$$r = \frac{s_{xy}}{s_x s_y}. \qquad (3.7)$$

As in the case of the population correlation coefficient, r also ranges from -1 to $+1$. Note that the sample covariance can be expressed as $s_{xy} = r s_x s_y$. As will be seen in the next section, s_{xy} is unbiased for S_{xy}. However, r in (3.7) is biased for ρ.

If units (U_1, U_4) appear in a sample of size two from the population in Table 2.1, the sample means of the verbal and math scores are $\bar{x} = (520 + 580)/2 = 550$ and $\bar{y} = (670 + 720)/2 = 695$. The sample variances and covariance are $s_x^2 = [(520 - 550)^2 + (580 - 550)^2]/1 = 1800$, $s_y^2 = [(670 - 695)^2 + (720 - 695)^2]/1 = 1350$, and $s_{xy} = [(520 - 550)(670 - 695) + (580 - 550)(720 - 695)]/1 = 1500$. From (3.7), the correlation is $r = 1500/(1800 \times 1350)^{1/2} = 0.96$. For this sample, the correlation is much larger than the population correlation of 0.41.

Grouped data

For the 50 states plus the District of Columbia in the U.S., the number of persons (y) over 25 years (in millions) and the percent (x) of the population with four or more years of college are grouped in Table 3.1 into a joint distribution and presented in Figure 3.1.

One can denote the mid-values of x by x_i, and its marginal frequencies by f_i, $i = (1, 2, ...)$, $\Sigma f_i = N$. The mean and variance of x now are obtained from

$$\bar{X} = \sum x_i f_i / N$$

Table 3.1. Joint distribution of age and education in 1995 in the 50 states and DC.

Percent Graduates (x)	Persons (millions) over 25 Years of Age (y)			
	0–2	2–4	4–20	Total
14–18	3	3	0	6
18–22	9	5	4	18
22–26	6	3	5	14
26–30	5	2	2	9
30–34	1	2	1	4
Total	24	15	12	51

Source: The U.S. Bureau of the Census, Current Population Survey, March 1996, and the *New York Times Almanac,* 1998, p. 374.

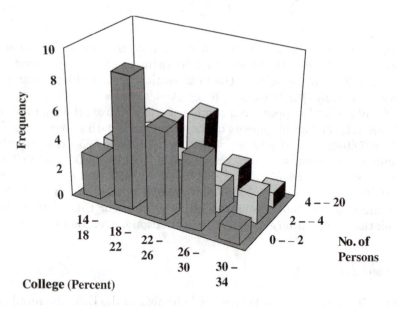

Figure 3.1. Joint distribution of the number of persons over 25 years of age and the percentage of college graduates in the 50 states and DC.

and

$$S_x^2 = \sum (x_i - \bar{X})^2 f_i / (N - 1). \tag{3.8}$$

The numerator of S_x^2 can be expressed as $\sum x_i^2 f_i - N\bar{X}^2$. For y, the mean $\bar{Y}$ and variance S_y^2 are obtained similarly. With the joint frequencies f_{ij},

Table 3.2. Arrangement of 40 population
units into four systematic samples.

Samples			
1	2	3	4
1	2	3	4
5	6	7	8
⋮	⋮	⋮	⋮
37	38	39	40

the covariance of x and y is obtained from

$$S_{xy} = \sum\sum (x_i - \bar{X})(y_i - \bar{Y})f_{ij}/(N-1) \qquad (3.9)$$

The numerator of this covariance can be expressed as $\sum\sum x_i y_i f_{ij} - N\bar{X}\bar{Y}$.
For the data in Table 3.1, $\bar{X} = (16 \times 6 + 20 \times 18 + \cdots + 32 \times 4)/51 = 22.98$, $S_x^2 = [(16 - 22.98)^2 \times 6 + (20 - 22.98)^2 \times 18 + \cdots + (32 - 22.98)^2 \times 4]/50 = 20.38$, and $S_x = (20.38)^{1/2} = 4.51$. Similarly, $\bar{Y} = 4.18$, $S_y^2 = 19.95$, and $S_y = 4.47$. From Equation 3.9, $S_{xy} = [(16 - 22.98) \times (1 - 4.18) \times 3 + (16 - 22.98)(3 - 4.18) \times 3 + \cdots + (32 - 22.98)(12 - 4.18) \times 1]/50 = 2.50$. Now, $r = 2.50/(4.51 \times 4.47) = 0.12$.
The joint probability distribution of two random variables and related properties are presented in Appendix A3.

3.5 Linear combination of two characteristics

It is frequently of interest to estimate the mean or total of a sum, difference, or a specified linear combination of two characteristics, for example, the math and verbal scores. Linear combinations of incomes of couples, scores of students on an aptitude test taken on two occasions, and sales or profits in two successive years provide other illustrations.

Linear combination

For each of the N population units, consider a linear combination of x_i and y_i, $g_i = ax_i + by_i + c$, where a, b, and c are chosen constants. This type of linear combination is of importance in several practical situations.

For example, the applicants for admission to a college may be ranked by an index of the SAT verbal and math scores with $a = 5/8$, $b = 3/8$, and $c = -500$.

The population total and mean of g_i are $G = aX + bY + c$ and $\bar{G} = (a\Sigma_1^N x_i + b\Sigma_1^N y_i + Nc)/N = a\bar{X} + b\bar{Y} + c$. As shown in Appendix A3, the variance of g_i is given by

$$S_g^2 = V(ax_i + by_i) = a^2 S_x^2 + b^2 S_y^2 + 2ab S_{xy}. \qquad (3.10)$$

From a simple of observations (x_i, y_i), $i = 1, 2, ..., n$, the sample mean of g_i is $\bar{g} = a\bar{x} + b\bar{y} + c$, and its variance is given by

$$s_g^2 = a^2 s_x^2 + b^2 s_y^2 + 2ab s_{xy}. \qquad (3.11)$$

As in the case of $\bar{x}$ and $\bar{y}$, for simple random sampling, $\bar{g}$ and s_g^2 are unbiased for $\bar{G}$ and S_g^2, respectively. Now, $E(s_g^2) = a^2 E(s_x^2) + b^2 E(s_y^2) + 2ab E(s_{xy})$, which is the same as S_g^2. Equating this expression to the right-hand side of (3.10), yields $E(s_{xy}) = S_{xy}$, that is, the sample covariance s_{xy} is unbiased for the population covariance S_{xy}. This result can also be obtained directly from the expression of s_{xy}.

Further, the variance of $\bar{g}$ is

$$V(\bar{g}) = (1 - f) s_g^2/n, \qquad (3.12)$$

and an unbiased estimator of this variance is obtained from $v(\bar{g}) = (1 - f)s_g^2/n$.

For the above index of the SAT scores, from Table 2.1, $\bar{G} = (5/8)(550) + (3/8)(670) - 500 = 95$, and $S_g^2 = (5/8)^2 (5840) + (3/8)^2 (3680) + 2(5/8) \times (3/8)(1920) = 3698.75$.

If the first and fourth units are selected into a sample of size two, $\bar{g} = (5/8)(550) + (3/8)(695) - 500 = 104.375$, and $s_g^2 = (5/8)^2 (1800) + (3/8)^2 (1350) + 2(5/8)(3/8)(1500) = 1596.09$.

From (3.12), $V(\bar{g}) = (6 - 2)(3698.75)/12 = 1232.92$ and hence S.E. $(\bar{g}) = 35.11$. The sample estimate of this variance is $(6 - 2)(1596.09)/12 = 532.03$, and hence the sample S.E. of $\bar{g}$ is equal to 23.07.

Note that all the sample observations are not needed to find $\bar{g}$ since it is the same as $a\bar{x} + b\bar{y} + c$. Similarly, the S.E. of $\bar{g}$ can be obtained by first finding s_g^2 from (3.11) as above.

The sum

The population and sample means and variances for the sum $t_i = x_i + y_i$ can be obtained directly or from the corresponding results for g_i in (3.10) with $a = 1$, $b = 1$, and $c = 0$. The total and mean of t_i are $T = X + Y$ and $\bar{T} = \bar{X} + \bar{Y}$. The population variance of t_i is

$$S_t^2 = \sum_1^N (t_i - \bar{T})^2/(N-1) = S_x^2 + S_y^2 + 2S_{xy}. \qquad (3.13)$$

The sample mean $\bar{t} = \bar{x} + \bar{y}$ is unbiased for $\bar{T} = \bar{X} + \bar{Y}$. Its variance is given by $V(\bar{t}) = (1-f)S_t^2/n$. An unbiased estimator of this variance is $v(\bar{t}) = (1-f)s_t^2/n$, where

$$s_t^2 = \sum_1^n (t_i - \bar{t})^2/(n-1) = s_x^2 + s_y^2 + 2s_{xy}. \qquad (3.14)$$

For the sum of the verbal and math scores in Table 2.1, the mean is 1220. If the first and fourth units are selected in a sample of size two, $\bar{t} = (550 + 695) = 1245$. From (3.13), $S_t^2 = 5840 + 3680 + 2(1920) = 13{,}360$. Hence, $V(\bar{t}) = 13{,}360/3 = 4453.33$ and S.E.$(\bar{t}) = 66.73$.

With the first and fourth units in the sample, the variance in (3.14) is $s_t^2 = 1800 + 1350 + 2(1500) = 6150$. Now, the sample estimate of the variance is $v(\bar{t}) = 6150/3 = 2050$, and hence the sample S.E. of $\bar{t}$ is equal to 45.28. This sample estimate is much smaller than the actual standard error of 66.73.

Note that S_{xy} in (3.13) can be replaced by $\rho S_x S_y$, and s_{xy} in (3.14) by $rs_x s_y$.

The variances of t_i and $\bar{t}$ will be larger when x and y are positively correlated than when they are uncorrelated or negatively correlated.

The difference

Denote the difference of the two characteristics by $d_i = x_i - y_i$, $i = 1$, $2, ..., N$. An unbiased estimator for the mean $\bar{D} = \bar{X} - \bar{Y}$ is $\bar{d} = \bar{x} - \bar{y}$. The population variance of this difference is

$$S_d^2 = \sum_1^N (d_i - \bar{D})^2/(N-1) = S_x^2 + S_y^2 - 2S_{xy}. \qquad (3.15)$$

and its unbiased estimator is

$$s_d^2 = \sum_1^n (d_i - \bar{d})^2/(n-1) = s_x^2 + s_y^2 - 2s_{xy}. \qquad (3.16)$$

The variance of $\bar{d}$ is $V(\bar{d}) = (1-f)S_d^2/n$, which is estimated from $v(\bar{d}) = (1-f)s_d^2/n$.

For the population in Table 2.1, $\bar{D} = 550 - 670 = -120$ and the variance in (3.15) is $S_d^2 = 5840 + 3680 - 2(1920) = 5680$. For samples of size two, the variance of $\bar{d}$ is $5680/3 = 1893.33$, and hence it has a standard error of 43.51.

If units (U_1, U_4) appear in the sample, $\bar{d} = 550 - 695 = -145$, and $s_d^2 = 1800 + 1350 - 2(1500) = 150$. The sample variance of $\bar{d}$ is $150/3 = 50$, and hence the S.E. of $\bar{d}$ becomes 7.1. This sample estimate is considerably smaller than the actual standard error of 43.51.

The variances of d_i and $\bar{d}$ will be larger when x and y are negatively correlated than when they are uncorrelated or positively correlated.

Covariance and correlation of the sample means

Note that

$$V(\bar{t}) = V(\bar{x} + \bar{y}) = V(\bar{x}) + V(\bar{y}) + 2 \, \text{Cov} \, (\bar{x}, \bar{y}). \qquad (3.17)$$

This variance however is the same as $V(\bar{t}) = (1-f)S_t^2/n = (1-f) \times (S_x^2 + S_y^2 + 2S_{xy})/n$. Thus, $\text{Cov}(\bar{x}, \bar{y}) = (1-f)S_{xy}/n$, which is estimated from $(1-f)s_{xy}/n$. This result can also be obtained from the expressions for the variance of $\bar{g}$ and its estimate described in Section 3.5 or of $\bar{d}$ in Section 3.5. When $n = 2$, for the math and verbal scores in Table 2.1, $\text{Cov}(\bar{x}, \bar{y}) = 1920/3 = 640$. If the first and fourth units appear in the sample, an estimate of this covariance is $1500/3 = 500$.

The correlation of the sample means is $\text{Cov}(\bar{x}, \bar{y})/[V(\bar{x})V(\bar{y})]^{1/2}$, which is the same as ρ.

3.6   Systematic sampling

Sample selection

In the common practice for selecting a systematic sample, the population is divided into k groups of size $n = N/k$ in each. One unit is chosen randomly from the first k units and every kth unit following it is included in the sample. If r is the random number drawn from the first group, units numbered $r + uk$, $u = (0, 1,..., n - 1)$ constitute the sample. This procedure results in the **1 in k systematic sample**.

Table 3.2 presents $N = 40$ units arranged in $k = 4$ groups of $n = 10$ units in each. If the random number selected from the first four numbers is 2, for example, the sample would consist of the units $(2, 6, 10, ..., 38)$. Selecting a systematic sample as above is equivalent to selecting randomly one of the k groups.

If N/k is not an integer, the sizes for the initial and latter samples will be equal to $[N/k] + 1$ and $[N/k]$, where $[N/k]$ is the integer closest to N/k. For instance, if $N = 39$ and $k = 4$, the first three samples will have size 10 and the last one, 9.

It is very convenient to draw systematic samples from telephone and city directories, automobile registries, and electoral rolls. A sample of households from a residential neighborhood can be easily obtained by this method. It also becomes convenient to draw a sample systematically when the population frame is not available. For example, selection of every 10th farm on the spot from the approximately 500 farms in a village will provide a sample of 50 farms. In industrial quality control, a sample of items produced, for example, every 30 minutes or every tenth batch of manufactured items is inspected. In some marketing and political surveys, interviews are attempted from, for example, every tenth person passing a certain location. For surveys supplementing censuses, systematic sampling is frequently employed. In large-scale multistage surveys, samples are selected systematically at the different stages.

Expectation and variance

For the systematic sampling procedure, the observations of the population can be denoted by y_{ij}, $i = 1,..., k$ and $j = 1,..., n$. Since each of the k samples has a $1/k$ chance of being selected, the expectation of the sample mean $\bar{y}_{sy} = \bar{y}_i$ is $\bar{Y}$ and hence it is unbiased. Its variance is given by

$$V(\bar{y}_{sy}) = \frac{1}{k} \sum_{1}^{k} (\bar{y}_i - \bar{Y})^2, \qquad (3.18)$$

which will be small if the variation among $\bar{y}_i$ is not large. Preliminary information on y_{ij} can be used to determine k and n for reducing this variance.

Since only $_nC_2 = n(n - 1)/2$ of the $_NC_2 = N(N - 1)/2$ pairs of the population units are given an equal chance for being selected, an unbiased

Table 3.3. Systematic samples of the savings
($1000) of 40 employees.

	\multicolumn{4}{c}{Samples}			
	1	2	3	4
	23	30	19	38
	23	38	19	18
	43	23	14	17
	40	17	40	23
	20	20	24	14
	14	22	24	14
	17	12	30	14
	20	14	15	14
	15	29	10	10
	16	22	16	12
Means	23.1	22.7	21.1	17.4
Variances	104.1	62.01	77.66	65.16

estimator of the variance in (3.18) cannot be found. Approximate procedures for this variance are available in the literature; see Cochran (1977, Chap. 8) and Section 3.9.4 below.

Example 3.1. Systematic sampling to estimate employee savings: The savings ($10,000) of 40 employees are arranged in Table 3.3 into $k = 4$ groups of size $n = 10$ in each. In actual sampling, one of the four samples will be selected randomly.

The population mean and variance of the savings are $\bar{Y} = 21.075$ and $S^2 = 76.48$. From (3.18), $V(\bar{y}_{sy}) = [(23.1 - 21.075)^2 + (22.7 - 21.075)^2 + (21.1 - 21.075)^2 + (17.4 - 21.075)^2]/4 = 5.062$, and hence S.E.$(\bar{y}_{sy}) = 2.25$. For simple random sampling, $V(\bar{y}) = (1 - 0.25)(76.48)/10 = 5.736$, and hence S.E.$(\bar{y}) = 2.4$. The gain in precision for $\bar{y}_{sy}$ relative to $\bar{y}$ is $(5.736 - 5.062)/5.062 = 0.13$ or 13%.

An unbiased estimator for the population total is obtained from $\hat{Y} = N\bar{y}_{sy}$, and S.E.$(\hat{Y}) = N$ S.E.$(\bar{y}_{sy})$. For the above example, the standard errors of $\hat{Y}$ for systematic and simple random sampling, respectively, are 40(2.25) = 90 and 40(2.4) = 96.

Comparison with simple random sampling

The variance within the systematic samples is

$$S_{wsy}^2 = \frac{\sum_1^k \sum_1^n (y_{ij} - \bar{y}_i)^2}{k(n-1)} = \frac{1}{k} \sum_1^k s_i^2. \tag{3.19}$$

Note that the **within-variance** s_i^2 of the ith sample has $(n - 1)$ d.f., and (3.19) is obtained by pooling the k sample variances.

From the expression for S^2 in Appendix A3,

$$V(\bar{y}_{sy}) = \frac{N-1}{N}S^2 - \frac{(n-1)}{n}S_{wsy}^2. \tag{3.20}$$

This variance decreases as the within-variance increases. If the units in a systematic sample vary, they represent the population units adequately. If n increases, the multiplier $-(n - 1)/n$ decreases but S_{wsy}^2 may increase. As a result, the variance in (3.20) need not decrease with an increase in the sample size; see Exercise 3.8.

From Equation 3.20, the difference in the variances of the mean $\bar{y}$ of a simple random sample and $\bar{y}_{sy}$ is

$$V(\bar{y}) - V(\bar{y}_{sy}) = \left(\frac{N-n}{Nn} - \frac{N-1}{N}\right)S^2 + \frac{(n-1)}{n}S_{wsy}^2$$

$$= \frac{n-1}{n}(S_{wsy}^2 - S^2). \tag{3.21}$$

Hence, $\bar{y}_{sy}$ will have higher precision than $\bar{y}$, if the within variance S_{wsy}^2 is larger than the population variance S^2.

For the four samples in Example 3.1, the within variances s_i^2 equal 104.1, 62.01, 77.66, and 65.16 as presented in Table 3.3. Hence, $S_{wsy}^2 = 77.23$, which is only slightly larger than the population variance $S^2 = 76.48$. This was the reason for the gain in precision for the mean of the systematic sample relative to the mean of the simple random sample to be as small as 13%.

Further properties

As noted in the above two sections, the precision of the systematic sample mean increases if $\bar{y}_i$ do not vary much but s_i^2 differ from each other. These two objectives can be met through any prior information available on the population units.

If the observations of the population units are in an ascending or descending order, that is, if there is a linear trend, Cochran (1977, pp. 215–216) shows that $V(\bar{y}_{sy})$ is smaller than $V(\bar{y})$. If the observations, however, are in a random order, both the procedures can be expected to have almost the same precision. In this case, the variance in (3.18) can be estimated from $v(\bar{y}) = (1 - f)\, s^2/n$ in (2.14). For the random ordering,

Madow and Madow (1944) show that the averages of $V(\bar{y}_{sy})$ and $V(\bar{y})$ over the $N!$ permutations of the observations are the same.

3.7   Nonresponse

In almost every survey, some of the units selected in the sample do not respond to the entire survey or to some of the items in the questionnaire. The different reasons for nonresponse and the various procedures for compensating for its effects are described in detail in Chapter 11. This section briefly examines the bias and MSE arising from nonresponse.

To analyze this situation, consider the population of size N to consist of the responding and nonresponding groups of sizes N_1 and $N_2 = N - N_1$, respectively. Denote the totals, means, and variances of these groups by (Y_1, Y_2), $(\bar{Y}_1, \bar{Y}_2)$, and (S_1^2, S_2^2), respectively. The total of the N population units is $Y = Y_1 + Y_2$. The population mean can be expressed as

$$\bar{Y} = Y/N = (Y_1 + Y_2)/N$$
$$= (N_1\bar{Y}_1 + N_2\bar{Y}_2)/N = W_1\bar{Y}_1 + W_2\bar{Y}_2, \qquad (3.22)$$

where $W_1 = N_1/N$ and $W_2 = N_2/N = 1 - W_1$ are the proportions of the units in the two groups.

In a sample of size n selected randomly from the N units, n_1 responses and $n_2 = n - n_1$ nonresponses will be obtained. The probability of (n_1, n_2) is given by

$$P(n_1, n_2) = (_{N_1}C_{n_1})(_{N_2}C_{n_2})/(_NC_n). \qquad (3.23)$$

Denote the sample mean and variance of the n_1 responses by $\bar{y}_1$ and s_1^2. Note that for the n_2 nonrespondents, the sample mean and variance, $\bar{y}_2$ and s_2^2, are not available.

The sample mean $\bar{y}_1$ is unbiased for $\bar{Y}_1$. However, for estimating the population mean, it has a bias of

$$B(\bar{y}_1) = E(\bar{y}_1) - \bar{Y} = \bar{Y}_1 - \bar{Y}$$
$$= \bar{Y}_1 - (W_1\bar{Y}_1 + W_2\bar{Y}_2) = W_2(\bar{Y}_1 - \bar{Y}_2). \qquad (3.24)$$

This **bias** is positive if $\bar{Y}_1$ is larger than $\bar{Y}_2$, and negative otherwise.

The absolute value of this bias depends on the difference of these means and on the size N_2 of the nonresponse group, but not on the sample size n.

The variance of $\bar{y}_1$ is $V(\bar{y}_1) = (1 - f_1)\, S_1^2/n_1$, and its MSE is given by

$$\mathrm{MSE}(\bar{y}_1) = V(\bar{y}_1) + B^2(\bar{y}_1). \qquad (3.25)$$

The number of respondents n_1 may increase with an increase of the sample size n. As a result, $V(\bar{y}_1)$ and $\mathrm{MSE}(\bar{y}_1)$ may decrease.

3.8   Inference regarding the means

Statistical tests of hypotheses available for large populations can also be used as approximate procedures for sampling from finite populations, unless the population and sample sizes are too small. The following subsections examine these tests for the mean of a population and the difference of the means of two populations.

Population mean

Consider the statement that the $N = 2000$ freshman and sophomore (FS) students in a college use the campus computer facilities on an average at least 6 hours a week. One can examine this statement from a sample of, for example, $n = 25$ students. For these selected students, consider the sample mean $\bar{y} = 8$ and variance $s^2 = 15$. Now, $v(\bar{y}) = [(2000 - 25)/2000](15/25) = 0.5925$, and hence S.E.$(\bar{y}) = 0.77$, which is almost 10% of the sample mean.

One may examine the null hypothesis, $\bar{Y} \le 6$ against the **alternative hypothesis**, $\bar{Y} > 6$, assuming that approximately $Z = (\bar{y} - \bar{Y})/\text{S.E.}(\bar{y})$ follows the standard normal distribution. Since $Z = 1.65$ for the significance level $\alpha = 0.05$, the 95% one-sided lower confidence limit for $\bar{Y}$ is given by $\bar{y} - 1.65\,\text{S.E.}(\bar{y}) \le \bar{Y}$, that is, $\bar{Y} \ge 8 - 1.65(0.77) = 6.73$. From this result, the null hypothesis is rejected at the 5% level of significance, and credence is given to the statement that the FS group utilizes the computer facilities for at least 6 hours a week.

Alternatively, with the null hypothesis value for $\bar{Y}$, one can compute $Z = (\bar{y} - \bar{Y})/\text{S.E.}(\bar{y})$, which is equal to $(8 - 6)/0.77 = 2.6$. Since this value exceeds 1.65, the null hypothesis is rejected at $\alpha = 0.05$ and the same inference as above is reached.

One can also formally examine the above hypotheses from finding the **p-value**, which is the probability of observing a sample mean of 8

or more when the null hypothesis is true, that is, $\bar{Y} \leq 6$. This probability is the same as $\Pr(Z \geq 2.6)$, which is equal to 0.0047 from the
tables of the standardized normal distribution. Since this probability
is small, the null hypothesis is rejected.

Since the population size is not too small, one may examine the
above hypotheses through the t-distribution. From the tables of this
distribution $t_{24} = 1.7109$ for $\alpha = 0.05$. The lower confidence limit for
$\bar{Y}$ is $8 - 1.7109(0.77) = 6.68$. Also, $t = (8 - 6)/(0.77) = 2.6$ exceeds
1.7109. The p-value now is $\Pr(t_{24} > 2.6)$, which is close to 0.0082. Each
of these three procedures comes to the same conclusion as above with
normal approximation.

Difference of two population means

Continuing with the illustration in the last section, consider the statement that the average number of hours of usage of the computer
facilities for the FS group is different from that of the 1800 junior and
senior (JS) students. With the subscripts 1 and 2 representing the FS
and JS groups, one can denote their means and variances by $(\bar{Y}_1, \bar{Y}_2)$
and (S_1^2, S_2^2). To examine the above statement, the null and alternative
hypotheses, respectively, are $\bar{Y}_1 = \bar{Y}_2$ and $\bar{Y}_1 \neq \bar{Y}_2$.

With $D = \bar{Y}_1 - \bar{Y}_2$, one can express these hypotheses as $D = 0$ and
$D \neq 0$. Notice that the alternative hypothesis is **two sided**, whereas
it was **one sided** in the last section.

For the $N_1 = 2000$ students of the FS group, a sample of size $n_1 =$
25 with $\bar{y}_1 = 8$ and $s_1^2 = 15$ has been considered. For the JS group,
consider an independent sample of size $n_2 = 20$ from the $N_2 = 1800$
students with mean $\bar{y}_2 = 10$ and variance $s_2^2 = 12$. The estimate of
$V(\bar{y}_2)$ is $v(\bar{y}_2) = [(1800 - 20)/1800](12/20) = 0.5933$.

From the above two samples, an estimate of D is $d = \bar{y}_1 - \bar{y}_2 =$
-2. Since $V(d) = V(\bar{y}_1) + V(\bar{y}_2)$, its estimate is $v(d) = v(\bar{y}_1) + v(\bar{y}_2) =$
$0.5925 + 0.5933 = 1.1858$, and hence S.E.$(d) = 1.09$. As an approxi-
mation, $Z = (d - D)/\text{S.E.}(d)$ follows the standard normal distribution.
Now, 95% lower and upper confidence limits for D are $-2 - 1.96(1.09) =$
-4.14 and $-2 + 1.96 (1.09) = 0.14$. Since these limits enclose zero,
the null hypothesis value for D, at the 5% level of significance, the
null hypothesis that the average number of hours of usage is the same
for both the FS and JS groups is not rejected.

Alternatively, $Z = -2/1.09 = -1.83$, which is larger than -1.96.
Hence at the 5% level of significance, the null hypothesis is again not

rejected. From the tables of Z, the p-value is equal to 0.0672, which may be considered to be large to reject the null hypothesis.

If both the populations are considered to be large with the same variance, the pooled estimate of the variance is obtained from $s^2 = [(n_1 - 1) \times s_1^2 + (n_2 - 1)s_2^2]/(n_1 + n_2 - 2)$ d.f. with $(n_1 + n_2 - 2)$ d.f. In this case, the variance of d is obtained from $v(d) = s^2(1/n_1 + 1/n_2)$, and the S.E.$(d)$ is obtained from the square root of this variance. Now, $t = d/$S.E.(d) follows the t-distribution with $(n_1 + n_2 - 2)$ d.f, and the above procedures are followed with this approximation.

Correlated samples

Now consider the statement that after improving the computer facilities, the average usage for the FS group increased by at least 2 hours a week.

Denote the number of hours of usage for the $N = 2000$ students initially by y_{1i} and after the improvements by y_{2i}, $i = 1, 2, ..., N$. With this notation, the mean and variance initially are $\bar{Y}_1$ and S_1^2. Similarly, the mean and variance after the improvements are $\bar{Y}_2$ and S_2^2. The increase for each student is $d_i = y_{2i} - y_{1i}$, and for the N students the mean of this difference is $D = \bar{Y}_2 - \bar{Y}_1$. The variance of d_i is $S_d^2 = S_2^2 + S_1^2 - 2S_{12}$.

The null and alternative hypotheses now are $D \le 2$ and $D > 2$. For the $N = 2000$ students of the FS group, a sample of size $n = 25$ with the mean and variance $\bar{y}_1 = 8$ and $s_1^2 = 15$ has been considered. For the same sample, after the improvements, consider the mean $\bar{y}_2 = 11.5$ and variance $s_2^2 = 18$, and the covariance $s_{12} = 12$ for the observations before and after the improvements. From these figures, the estimate of D is given by $d = \bar{y}_2 - \bar{y}_1 = 3.5$, and the estimate of S_d^2 by $s_d^2 = s_2^2 + s_1^2 - 2s_{12} = 11$. The variance of d is $V(d) = (1 - f)S_d^2/n$, where $f = n/N$. From the sample observations, an estimate of this variance is $v(d) = (1 - f)s_d^2/n = (79/80)11/25 = 0.4345$, and hence S.E.$(d) = 0.66$.

Now, with $\alpha = 0.05$, the one-sided lower confidence limit for D is given by $D \ge 3.5 - 1.65 (0.66) = 2.41$. Hence, it may be inferred that for the FS group, the average usage of the computer facilities has increased at least by 2 hours a week. Since $Z = (d - 2)/$S.E.$(d) = 2.3$, which exceeds 1.65, the same conclusion is reached from this standardized value. From the tables of Z, the p-value in this case is close to 0.01.

The t-distribution in this case will have 24 d.f. as in Section 3.8, and the p-value now is close to 0.0168.

3.9   Sampling with replacement

If a sample of n units is selected randomly and **with replacement** from the population, any unit U_i, $i = (1, 2, ..., N)$, may appear more than once in the sample. This procedure is not practical for sampling from a finite population. However, one can compare its properties with sampling randomly without replacement, presented in Section 2.5.

For this procedure, there are N^n possible samples. The probabilities of selection of the units follow the binomial distribution with n trials and the probability of selection of a unit equal to $1/N$ at any trial. The sample mean $\bar{y}$ is unbiased for $\bar{Y}$ as in the case of the without replacement procedure, but $V(\bar{y}) = \sigma^2/n = (N - 1)S^2/Nn$. An unbiased estimator of this variance is $v(\bar{y}) = s^2/n$; see Exercises 3.19 and 3.20.

Notice that the variance of $\bar{y}$ for this method is larger than the variance in (2.12) for the without replacement procedure. The difference between these procedures becomes small if the population size is large. By deleting the duplications, the sample mean obtained from the **distinct** units will be unbiased for $\bar{Y}$, and its variance will be smaller than $(N - 1)S^2/Nn$.

Exercises

3.1.   To estimate the mean of a population of 3000 units, consider increasing the sample size from 5 to 10%. For the sample mean, find the relative (a) decrease in the variance, (b) decrease in the standard error, and (c) decrease in the coefficient of variation and (d) increase in the precision. (e) What will be the relative decrease in the confidence width for a specified probability? (f) What will be the changes in (a) through (e) for the estimation of the population total?

3.2.   (a) Find the means, variances, covariance, and correlation of the 30 verbal and math scores in Table T2 in the Appendix. (b) Group the data into a bivariate distribution with classes 350 to 470, 470 to 590, and 590 to 710 for the verbal scores and 400 to 520, 520 to 640, and 640 to 760 for the math scores. Find the means, variances, covariance, and correlation from this distribution, and compare them with the results in (a).

3.3.   From the information in Exercise 2.1, for the difference in the means of the largest and middle-sized corporations, find (a) the estimate, (b) its S.E., and (c) 95% confidence limits. (d) Explain whether the procedure used for (a) provides an unbiased estimator.

3.4. From the information in Exercise 2.1, for the ratio of the means of the middle-sized to the smallest corporations, find (a) the estimate, (b) its S.E., and (c) the 95% confidence limits. (d) Explain whether the procedure used for (a) provides an unbiased estimator.

3.5. For a random sample of 20 from 2000 candidates, the averages and standard deviations of the SAT verbal scores at the first and second attempts were (520, 150) and (540, 160), respectively. The sample correlation coefficient of the scores on the two occasions was 0.6. For the increase in the average score, find (a) the estimate, (b) its S.E., and (c) the 95% confidence limits.

3.6. Discuss whether the following types of systematic sampling result in higher precision than simple random sampling. (a) Every fifth sentence to estimate the total number of words on a page. (b) Every tenth student entering the bookstore of a university from 12 to 2 P.M. on a Monday to estimate the average amount spent by all the students of that university for books. (c) Every fourth house on each street of a region consisting of one, two, three, or more member families to estimate the total number of children attending schools in the region. (d) Every tenth person leaving a polling booth between 6 and 9 P.M. to predict the winning candidate in a political election.

3.7. Three different types of lists are available for the 1200 employees of an educational institution. In the first one, they appear in alphabetical order. In the second list, they are arranged according to the years of employment. About 600 in the list are employed for 10 or more years, 400 from 5 to 9 years, and the rest for less than 5 years. In the third list, the names appear alphabetically, in succession for the above three groups. It is planned to estimate each of the following items with a 5% simple random or systematic sample: (a) annual savings, (b) number of children in the college, and (c) number of hours spent for sports and recreation during a typical week. For each case, recommend the lists and the type of sampling.

3.8. For the population of the savings in Table 3.3, find the means and S.E. values of $\bar{y}_{sy}$ for (a) one in three and (b) one in eight systematic samples. Compare the precisions of these means with the one in Example 3.1 for $k = 4$ and the means of simple random samples of equivalent sizes.

3.9. In the pilot survey of $n = 25$ selected randomly from the $N = 4000$ students mentioned in Section 1.8, only $n_1 = 15$ responded to the question on the number of hours (y_i) of utilizing the athletic facilities during a week. The mean and standard deviation of the responses were $\bar{y}_1 = 5$ and $s_1 = 3$. The mean $\bar{y}_2$ and standard deviation s_2 for the $n_2 = n - n_1 = 10$ nonrespondents are not known. For the population mean $\bar{Y}$ of the 4000 students, three estimates may be considered: (1) $\bar{y}_1$, (2) substitute $\bar{y}_1$ for each of the n_2 nonrespondents, and (3) substitute zero for each of the nonrespondents. (a) Find expressions to the biases, S.E.values and MSEs of these three estimating procedures. (b) Describe the conditions for which the biases and S.E. values can be large or small.

3.10. Blood pressures of a sample of 20 persons are presented in Table T6. Test the following hypotheses at a 5% level of significance in each case. (a) Before the treatment, the mean systolic pressure is at least 165. (b) Before the treatment, the mean diastolic pressure is at most 90. (c) After the treatment, the mean systolic pressure is 145. (d) After the treatment, the mean diastolic pressure is 75.

3.11. (a) From the sample in Table T6, find the 90% confidence limits for the mean systolic pressure before the treatment. (b) From these limits, is the same conclusion for the hypothesis reached as in Exercise 3.10(a).

3.12. From the sample in Table T6, test the hypotheses at the 5% level of significance that the treatment reduces the means of the systolic and diastolic pressures by 25 and 15, respectively.

3.13. *Project.* Find the covariance s_{xy}, the coefficient of variation $s_y/\bar{y}$, and the correlation coefficient r for each of the 20 samples selected in Exercise 2.10, and verify that s_{xy} is unbiased for S_{xy} but $s_y/\bar{y}$ is biased for the population coefficient of variation $S_y/\bar{Y}$ and r is biased for ρ. For r and s_{xy}, find the (a) bias, (b) variance, and (c) MSE.

3.14. It was shown in Section 3.5 that the sample covariance s_{xy} is unbiased for the population covariance S_{xy}. From the expressions in (3.4) and (3.6), show directly that s_{xy} is unbiased for S_{xy}.

3.15. From the Cauchy–Schwarz inequality in Appendix A3, show that (a) $-1 \le \rho \le 1$ and (b) $-1 \le r \le 1$.

3.16. Show that $1/\bar{y}$ is positively biased for $1/\bar{Y}$ by noting that (a) the arithmetic mean is larger than the harmonic mean, and (b) $\bar{y}$ and $1/\bar{y}$ are negatively correlated, and also (c) from the Cauchy–Schwartz inequality in Appendix A3.

3.17. (a) Show that

$$\sum_{i<j}^{N}\sum^{N}(x_i - x_j)(y_i - y_j) = N\sum_{1}^{N}(x_i - \bar{X})(y_i - \bar{Y}),$$

and hence S_{xy} in (3.4) is the same as

$$\sum_{i<j}^{N}\sum^{N}(x_1 - x_j)(y_i - y_j)/N(N-1).$$

(b) Similarly, show that s_{xy} in (3.6) can be expressed as

$$\sum_{i<j}^{n}\sum^{n}(x_i - x_j)(y_i - y_j)/n(n-1).$$

(c) From these expressions, show that for simple random sampling without replacement, s_{xy} is unbiased for S_{xy}.

3.18. For simple random sampling **without** replacement, show by using the properties in Section 2.5 that (a) $E(y_i) = \bar{Y}$, (b) $V(y_i) = (N-1)S^2/N = \sigma^2$, and (c) $\mathrm{Cov}(y_i, y_j) = -S^2/N$. With these results, derive the formula for $V(\bar{y})$ in (2.12).

3.19. For random sampling **with** replacement, show that (a) $E(y_i) = \bar{Y}$, (b) $V(y_i) = (N-1)S^2/N = \sigma^2$, and (c) $\mathrm{Cov}(y_i, y_j) = 0$. With these results, show that the sample mean $\bar{y}$ is unbiased for $\bar{Y}$ and its variance is given by $V(\bar{y}) = \sigma^2/n$. Note that this variance is larger than the variance for the without replacement procedure.

3.20. *Project.* Select all the $6^2 = 36$ samples of size two randomly with replacement from the six units in Table 2.1. Show that (a) $\bar{y}$ is unbiased for $\bar{Y}$ and its variance is given by

$(N - 1)S^2/Nn$, (b) s^2 is unbiased for σ^2 and (c) the mean of the distinct units in the sample is unbiased for $\bar{Y}$, and its variance is smaller than $(N - 1)S^2/Nn$.

Appendix A3

Joint distribution

Consider two random variables (X, Y) taking values (x_i, y_j) with probabilities $P(x_i, y_j)$, $i = 1, ..., k$; $j = 1, ..., l$, $\Sigma_1^k \Sigma_1^l P(x_i, y_j) = 1$. The probability for X taking the value x_i is $P(x_i) = \Sigma_j P(x_i, y_j)$, and the probability for Y taking the value y_j is $P(y_j) = \Sigma_i P(x_i, y_j)$. Further, $\Sigma_i P(x_i) = 1$ and $\Sigma_j P(y_j) = 1$.

The data in Table 3.1 provide an illustration. For example, the probability $P(x_1 = 16, y_1 = 1) = 3/51$. The random variable X takes the five values $(16, 20, 24, 28, 32)$, with probabilities $(6/51, 18/51, 14/51, 9/51, 4/51)$. The random variable Y takes the three values $(1, 3, 12)$ with probabilities $(24/51, 15/51, 12/51)$.

Expectation and variance

The **expected value** and **variance** of a discrete random variable X taking values x_i with probabilities $P(x_i)$, $i = 1, ..., k$, $\Sigma_1^k P(x_i) = 1$, are

$$E(X) = \sum_1^k x_i P(x_i)$$

and

$$V(X) = E[X - E(X)]^2 = \sum_1^k [x_1 - E(x_i)]^2 P(x_i).$$

The formula for the variance can also be expressed as

$$V(X) = E(X^2) - [E(X)]^2 = \left[\sum_1^k x_i^2 P(x_i)\right] - \left[\sum_1^k x_i P(x_i)\right]^2.$$

The above expressions are compact forms of the expectation and variance of a random variable presented in Appendix A2. The expectation and variance of the random variable Y are given by similar expressions.

Covariance and correlation

The **covariance** of X and Y is

$$\text{Cov}(X, Y) = E\{[X - E(X)][Y - E(Y)]\}$$

$$= \sum_{1}^{k}\sum_{1}^{l}[x_i - E(X)][Y_j - E(Y)]P(x_i, y_j).$$

This formula can also be expressed as

$$\text{Cov}(X, Y) = E(XY) - E(X)E(Y)$$

$$= \sum_{1}^{k}\sum_{1}^{l}x_iy_jP(x_i, y_j) - \left[\sum_{1}^{k}x_1P(x_i)\right]\left[\sum_{1}^{l}y_jP(y_j)\right].$$

The **correlation coefficient** of X and Y is

$$\rho = \frac{\text{Cov}(X, Y)}{\sqrt{V(X)}\sqrt{V(Y)}},$$

where $\sqrt{V(X)}$ and $\sqrt{V(Y)}$ are the **standard deviations** of X and Y. The range for ρ falls between -1 and 1.

Conditional and unconditional expectations and variances

In several instances, expectations and variances of the sample quantities for repetitions of the samples, for two or more stages of sampling, will be needed. The following results will be useful on such occasions.
    The expectation of Y, $\sum_{1}^{l}y_jP(y_j)$, can also be expressed as

$$E(Y) = \sum_{1}^{k}\sum_{1}^{l}y_jP(x_i, y_j)$$

$$= \sum_{1}^{k}P(x_i)\sum_{1}^{l}y_j[P(x_i, y_j)/P(x_i)]$$

$$= \sum_{1}^{k}P(x_i)\sum_{1}^{l}y_jP(y_j \mid x_i),$$

where $P(y_j | x_i)$ is the **conditional probability** of Y taking the value y_j when X takes the value x_i. The second summation on the right-hand side is the conditional mean of Y at $x = x_i$, which can be expressed as $E(Y | x_i)$, or as $E(Y | X)$. Thus, the above formula can be expressed as

$$E(Y) = E[E(Y | X)],$$

where the outer expectation on the right-hand side refers to the expectation with respect to X.

The variance of Y is

$$V(Y) = E(Y^2) - [E(Y)]^2.$$

Further, as in the case of the expectation of Y, $E(Y^2) = E[E(Y^2 | X)]$. Hence, $V(Y)$ can be expressed as

$$V(Y) = E[E(Y^2 | X)] - \{E[E(Y | X)]\}^2.$$

Subtracting and adding $E[E(Y|X)]^2$,

$$V(Y) = E\{E(Y^2 | X) - [E(Y | X)]^2\} + E[E(Y | X)]^2 - \{E[E(Y | X)]\}^2.$$

The expression in the first braces is the conditional variance of Y, and the third and fourth terms together are equal to the variance of $E(Y | X)$. Thus, $V(Y)$ can be expressed as

$$V(Y) = E[V(Y | X)] + V[E(Y | X)]$$

In this expression, the operators E and V inside and outside the square brackets refer to the expectation and variance conditional and unconditional on X, respectively.

Variance of a linear combination

$$
\begin{aligned}
S_g^2 &= V(ax_i + by_i) = \sum [(ax_i + by_i + c) - (a\bar{X} + b\bar{Y} + c)]^2 / (N - 1) \\
&= \sum [a(x_i - \bar{X}) + b(y_i - \bar{Y})]^2 / (N - 1) \\
&= a^2 \sum (x_i - \bar{X})^2 / (N - 1) + b^2 \sum (y_i - \bar{Y})^2 / (N - 1) \\
&\quad + 2ab \sum (x_i - \bar{X})(y_i - \bar{Y}) / (N - 1) \\
&= a^2 S_x^2 + b^2 S_y^2 + 2ab S_{xy}.
\end{aligned}
$$

An expression for S^2

The population variance S^2 can be expressed as

$$(N-1)S^2 = \sum_{1}^{k}\sum_{1}^{n}(y_{ij} - \bar{Y})^2 = \sum_{1}^{k}\sum_{1}^{n}[(y_{ij} - \bar{y}_i) + (\bar{y}_i - \bar{Y})]^2$$

$$= \sum_{1}^{k}\sum_{1}^{n}(y_{ij} - \bar{y}_i)^2 + n\sum_{1}^{k}(\bar{y}_i - \bar{Y})^2$$

$$= k(n-1)S_{wsy}^2 + nkV(\bar{y}_{sy}).$$

Cauchy–Schwartz inequality

For a set of numbers (a_i, b_i), $i = 1, 2, ..., k$,

$$\left(\sum_{1}^{k}a_i^2\right)\left(\sum_{1}^{k}b_i^2\right) \geq \left(\sum_{1}^{k}a_i b_i\right)^2,$$

and the equality occurs if a_i/b_i is a constant for $i = 1, 2, ..., k$.

CHAPTER 4

Proportions, Percentages, and Counts

4.1   Introduction

Employment rates, educational levels, economic status, and health conditions of the public are estimated at regular intervals through the surveys of national organizations. Marketing analysts estimate preferences of the public from the surveys on consumer panels. Private agencies obtain opinions of the public on issues such as economic situation, school budgets, health care programs, and changes in legislation. Polls are conducted to ascertain the choices of the public on political appointments and to predict outcomes of elections.

To estimate the proportions or percentages and the total numbers or counts of the population in the above types of categories or classes, respondents may be asked to provide a simple "Yes" or "No" answer to each of the survey questions. Similarly, responses to some types of questions may take the form of *approve* or *disapprove* and *agree* or *disagree*. Some surveys related to opinions and attitudes may request the respondents to specify their preferences on a scale usually ranging from zero or one to five or ten. The options *Uncertain, Undecided,* or *Don't know* are added to some questions in opinion surveys.

In this chapter, estimation of the population proportions and total counts for the case of two classes is described first, and then extended to three or more classes. Topics in the following sections also include confidence limits for the proportions and total counts, methods for finding sample sizes for specified criteria, and estimation of population sizes.

4.2   Two classes

Consider a qualitative characteristic with the Yes–No classification. Let C denote the number of the N population units falling into the first class. The proportion of the units in this class is

$$P = \frac{C}{N}. \tag{4.1}$$

The complementary group consists of $(N - C)$ units and their proportion is $Q = (N - C)/N = 1 - P$.

In a random sample of n units selected without replacement from the N units, denote the numbers of units falling into the two classes by c and $(n - c)$. The sample proportion of the units in the first group is

$$p = \frac{c}{n}. \tag{4.2}$$

The proportion in the complementary group is $q = (n - c)/n = 1 - p$.

Correspondence among the mean, total, and variance of a quantitative variable considered in Chapter 2 and the proportion and count of a qualitative characteristic can be established as follows. Assigning the numerals 1 or 0 to y_i if a unit belongs to the first or the second group, the total, mean, and variance become

$$Y = \sum_1^N y_i = C, \qquad \bar{Y} = \frac{C}{N} = P, \tag{4.3}$$

and

$$S^2 = \frac{\sum^N y_i^2 - N\bar{Y}^2}{N - 1} = \frac{C - NP^2}{N - 1} = \frac{N}{N - 1}PQ. \tag{4.4}$$

Similarly, with this **mnemonic** notation,

$$\bar{y} = \frac{\sum_1^n y_i}{n} = \frac{c}{n} = p \tag{4.5}$$

and

$$s^2 = \frac{\sum_1^N y_i^2 - n\bar{y}^2}{n - 1} = \frac{c - np^2}{n - 1} = \frac{n}{n - 1}pq. \tag{4.6}$$

Since $\bar{y}$ is unbiased for $\bar{Y}$, the sample proportion p is unbiased for P. From (2.12) and (4.4),

$$V(p) = \frac{N - n}{N - 1}\frac{PQ}{n}. \tag{4.7}$$

From (2.14) and (4.6), the estimator of this variance is

$$v(p) = \frac{N - n}{N} \frac{pq}{n - 1}.$$ (4.8)

The actual and estimate of the standard errors of p are obtained from the square roots of (4.7) and (4.8), respectively.

An unbiased estimator for Q is given by the sample proportion q. Its variance and estimator of variance are the same as (4.7) and (4.8), respectively. It is directly shown in Appendix A4 that the sample proportion and the variance in (4.8) are unbiased.

The probability of observing c and $(n - c)$ units in the sample from the C and $(N - C)$ units of the two groups is given by the **hypergeometric** distribution:

$$\Pr(c, n - c) = \binom{C}{c}\binom{N - C}{n - c} / \binom{N}{n}.$$ (4.9)

It is shown in Appendix A4 that for large N, (4.9) can be approximated by the **binomial** distribution:

$$\Pr(c) = \frac{n!}{c!(n - c)!} P^c Q^{n-c}.$$ (4.10)

For this distribution, $p = c/n$ is unbiased for P with $V(p) = PQ/n$, and an unbiased estimator of this variance is $v(p) = pq/n$. Note that for large N, (4.7) becomes the same as PQ/n. If n is also large, the difference of (4.8) from pq/n will be negligible.

Example 4.1. Hospital ownership: The American Hospital Association *Guide to Health Care Field* (1988) contains information on the hospitals regarding the number of physicians, occupancy rates, annual expenses, and related characteristics. In short-term hospitals, patients' stay is limited to 30 days. For medical and surgical services, New York State has 217 short-term hospitals. Of these hospitals, 150 are owned by individuals, partnerships, or churches. The remaining 67 are operated by federal, state, local, or city governments. In a random sample of 40 selected from the 217 hospitals, 29 were found to be in the first group and 11 in the second.

The estimate of the proportion of the hospitals in the first group is $p = 29/40 = 0.725$, or close to 73%. From (4.8),

$$v(p) = \frac{(217 - 40)}{217} \frac{(0.725)(0.275)}{39} = 0.0042$$

and S.E.$(p) = 0.0648$. From these figures, an estimate of the proportion
of the hospitals in the second group is 0.275, and it has the same S.E.

4.3   Total number or count

An unbiased estimator of the total number is $\hat{C} = Np$. Its variance
and estimator of variance are given by

$$V(\hat{C}) = N^2 V(p) = \frac{N^2(N-n)}{N-1} \frac{PQ}{n} \qquad (4.11)$$

and

$$v(\hat{C}) = N^2 v(p) = N(N-n)\frac{pq}{n-1}, \qquad (4.12)$$

respectively.

From the sample in Example 4.1, an estimate for the total number
of hospitals in the first group is $\hat{C} = 217(0.725) = 157$, and from (4.12),
S.E.$(\hat{C}) = 217(0.0648) = 14$.

4.4   Confidence limits

If N and n are large, $(p - P)/\sqrt{V(p)}$ approximately follows the standard normal distribution. For large N, this approximation becomes
increasingly valid as n is larger than 25 or 30 and P is close to 0.5.

With this approximation, the lower and upper limits for P are
obtained from

$$p \pm Z. \text{S.E.}(p), \qquad (4.13)$$

where S.E.(p) is obtained from (4.8) and Z as before is the $(1 - \alpha)$
percentile of the standard normal distribution. Since a discrete variable is being approximated by a continuous one, the *continuity correction* $1/2n$ can be added to the upper limit in (4.13) and subtracted from
the lower limit. If n is large, the effect of this correction becomes
negligible.

Example 4.2. Hospital ownership: For the illustration in Example 4.1, S.E.$(p) = 0.0648$. Approximate 95% confidence limits for the proportion in the first group are given by $0.725 \pm 1.96(0.0648)$; that is, the lower and upper limits are 0.598 and 0.852, respectively. The limits for the total number of hospitals in this group are $217(0.598) = 130$ and $217(0.852) = 185$.

4.5 Sample sizes for specified requirements

Similar to the procedures in Section 2.12 for the mean and total of a quantitative variable, sample sizes to estimate P or C can be found with specified requirements. The following types of specifications are usually required for estimating proportions in industrial research as well as for estimating percentages in political and public polls.

1. The variance of p should not exceed V; that is, S.E.(p) should not exceed $\sqrt{V}$. With this specification, from the expression for the variance in (4.7), the minimum sample size required is given by

$$n = \frac{n_l}{1 + (n_l - 1)/N}, \qquad (4.14)$$

where $n_l = PQ/V$ is the size needed for a large population.

For each of the following specifications, the sample size n_l for large N is presented, and the required sample size is obtained from (4.14).

2. The error of estimation $|p - P|$ should not exceed e, except for an α% chance. This specification is the same as

$$\Pr[|p - P| / \text{S.E.}(p) \le e / \text{S.E.}(p)] = 1 - \alpha. \qquad (4.15)$$

Assuming that $|p - P|/\text{S.E.}(p)$ follows the standard normal distribution, the right-hand side term inside the square brackets is the same as the standard normal deviate Z corresponding to the $(1 - \alpha)$ probability. Hence

$$n_l = Z^2 PQ / e^2. \qquad (4.16)$$

3. Confidence width for P, for a confidence probability $(1 - \alpha)$, should not exceed w. With the normal approximation, the width of the

interval in (4.13) is $2Z$ S.E.(p). Since it should not exceed w,

$$n_l = 4Z^2 PQ / w^2, \tag{4.17}$$

and n is obtained from (4.14).

4. The coefficient of variation of p, $\sqrt{V(p)}/P$, should not exceed C. In this case,

$$n_l = Q / PC^2. \tag{4.18}$$

5. The relative error $|p - P|/P$ should not exceed **a**, except for an $\alpha\%$ chance. For this requirement,

$$n_l = Z^2 Q / P\mathbf{a}^2. \tag{4.19}$$

The first three specifications require prior knowledge of the unknown proportion. In contrast, the last two specifications can be made without such information. The solutions to the sample sizes for all these five specifications, however, depend on the unknown proportions. If such information is not available, the sample sizes for the first three requirements can be found with $P = Q = 0.5$, since $V(p)$ takes its maximum in this case. For the remaining two specifications, relatively larger sample sizes are needed when P is between 0 and 0.5.

To estimate the percentages of qualitative characteristics, the proportion p, its S.E., and the specifications are expressed in percentages. The same solutions for the sample sizes are obtained again.

The first specification is equivalent to the requirement that the variance of $\hat{C}$ should not exceed $N^2 V$. Thus, n in (4.14) is also the sample size needed for estimating C with this requirement. Similarly, the sample sizes with the remaining four specifications satisfy the equivalent requirements for the estimation of C.

As in the case of the population mean and total, larger sample sizes are needed to estimate the proportions, percentages, and total numbers with smaller S.E. values, errors of estimation, confidence widths, or relative errors, and with larger confidence probabilities.

Example 4.3. Sample size: To estimate the proportion P of a large population having a given attribute, with an error not exceeding 5%

except for a 1 in 20 chance, from (4.16),

$$n_l = (1.96)^2 PQ/(0.05)^2.$$

If it is known that P is not too far from 0.5, $n_l = 384$ from this solution.
   If $P = 0.5$ and $\alpha = 0.05$ as above, and the relative error should not
exceed 1%, n_l from (4.19) is again equal to 384.
   With the above justifications, samples of size around 400 are frequently used in several political, public, and marketing surveys, even
when the population size is very large. Estimates of the proportions with
high precisions will be obtained, if the population is first divided into
homogeneous groups or strata and sample sizes for each group are
determined through the above types of requirements. With such a procedure, estimates for a population proportion are obtained by suitably
weighting the estimates from the different groups.

4.6   Political and public polls

A number of polls are conducted during election times to summarize
the preferences of the voters and to predict the winners. The results of
five of the 1992 presidential election polls are presented in Table 4.1.
These pre-election polls were conducted from October 28 to November 1
by the Gallup and Harris organizations, major newspapers, and television networks. Sample sizes for these surveys ranged from 982 to
1975. The last entry in the table is from an exit poll at 300 polling

Table 4.1. Presidential election—Percentages of voters for the three
candidates.

Description	Clinton	Bush	Perot
Pre-election polls			
$n = 9{,}115$	44	37	16
1,562	44	36	14
1,975	44	38	17
2,248	44	35	15
982	44	36	15
Election day			
$n = 15{,}490$	43	38	19

Source: The New York Times, November 4 and 5, 1993.

Table 4.2. Public surveys—Percentages of responses for different items.

Are you a better than average driver? $n = 597$	Better	Worse	Same
	49	3	40
Problem with today's movies $n = 597$	Sex	Violence	Both
	13	44	38
President Clinton's health-care program $n = 500$	Favor	Oppose	Not Sure
Sept. 23	57	31	12
Oct. 28	43	36	21

Source: Time Magazine of December 1992 for the first two surveys and of November, 1993 for the last two.

Table 4.3. Classification of voters' participation in the primaries on Super Tuesday, March 7, 2000; (percentages).

	Men	Women	No. of Years of College			
			<4	4	>4	None
Democrats	42	58	21	22	31	26
Republicans	55	45	26	24	20	30

Source: Voters News Service poll of 1698 Democrats and 1432 Republicans at the primary election places throughout New York State on Super Tuesday, March 7, 2000; reported in New York Times of March 9, 2000, page A20.

stations on a total of 15,490 voters, conducted on election day November 3, 1992 by Voters Research and Surveys.

Results of four surveys of general interest to the public conducted by Yankelovich Inc. are presented in Table 4.2. The first two surveys were conducted in December 1992, and the last two in September and October of 1993.

Classification of 1698 Democratic and 1432 Republican registered voters who participated in the primaries on Super Tuesday, March 7, 2000, is presented in Table 4.3.

For these surveys, the error of estimation $\mathbf{e} = |p - P|$, which is also referred to as the **sampling error** or **margin of error**, is computed for a confidence probability of 95% with the normal approximation; that is, as described in Section 4.5 for the second criterion, $\mathbf{e}$ is obtained from equating $\mathbf{e}/S.E.(p)$ to 1.96. The data from these tables will be analyzed in the following sections and in the exercises.

4.7   Estimation for more than two classes

In Tables 4.1 and 4.2, responses to the surveys are classified into three groups, and they do not add to 100% for some of the surveys. The remaining percentages of the persons selected in the samples probably refused to respond to the surveys or they were not sure of their responses. If the nonresponse group is included, the populations for these surveys can be considered to consist of four classes.

As presented in Table 4.2, in the sample of 500 persons contacted on September 23, the health-care program was favored by 57% and opposed by 31%. The remaining 12% were not sure of their preferences.

In Table 4.3, in addition to the Men–Women classification, voters are classified into four groups according to the number of years of college education.

Probabilities of the outcomes

To represent the above type of situations, let C_1, C_2, ..., C_k denote the number of units of a population belonging to k classes defined according to a qualitative characteristic. The corresponding proportions of the units are $P_1 = C_1/N$, $P_2 = C_2/N$, ..., $P_k = C_k/N$. The probability of observing c_1, c_2, ..., c_k units in the classes in a random sample of size n is

$$P(c_1, c_2, ..., c_k) = \binom{C_1}{c_1}\binom{C_2}{c_2} \cdots \binom{C_k}{c_k} / \binom{N}{n}. \qquad (4.20)$$

This distribution is an extension of (4.9) for $k \geq 3$ classes.

The estimates of the proportions and their variances can be obtained through the procedures in Section 4.2. The sample proportion $p_i = c_i/n$, $i = 1, 2, ..., k$, is unbiased for P_i. Its variance $V(p_i)$ is obtained from (4.7) by replacing P with P_i. Similarly, the estimator of variance $v(p_i)$ is obtained from (4.8) by replacing p with p_i. Unbiased estimator for the number of units in the ith class is $\hat{C}_i = Np_i$. The S.E. of $\hat{C}_i$ is given by $N\sqrt{V(p_i)}$ and it is estimated from $N\sqrt{v(p_i)}$. As in the case of two classes, confidence limits for the proportions and the numbers of units in the different classes are obtained by adding and subtracting Z times the respective S.E. values to the estimates.

Multinomial approximation

If N is large, the probability in (4.20) becomes

$$P(c_1, c_2, ..., c_k) = \frac{n!}{c_1! c_2! ... c_k!} P_1^{c_1} P_2^{c_2} ... P_k^{c_k}, \qquad (4.21)$$

which is the **multinomial** distribution. This expression refers to the chance of observing c_i units in the ith class in n independent trials, with the probabilities P_i remaining constant from trial to trial. For this distribution, $p_i = a_i/n$ is unbiased for P_i. Its variance and estimator of variance are $V(p_i) = P_i(1-P_i)/n$ and $v(p_i) = p_i(1 - p_i)/n$.

> **Example 4.4.** Presidential candidates: The sample size for the fourth survey in Table 4.1 is relatively large. Observations from this survey can be used for estimating the standard errors of the sample proportions and to find confidence intervals for the proportions of the population of voters favoring each of the three candidates.
>
> For Mr. Clinton, an estimate of the proportion is 0.44. Its sample variance is $(0.44)(0.56)/2247 = 0.0001$ and hence the S.E. of the sample proportion is 0.01. The 95% confidence limits for the population proportion of the voters favoring him are $0.44 - 1.96(0.01) = 0.42$ and $0.44 + 1.96(0.01) = 0.46$. Thus, an estimate for the percentage of the voters favoring Mr. Clinton is 44, with the 95% confidence interval of 42 to 46%.
>
> Similarly, 35% of the voters are for Mr. Bush, and the confidence limits are (33, 37)%. For Mr. Perot, the estimate is 15% and the confidence limits are (14, 16)%.

4.8   Combining estimates from surveys

The presidential election is an example in which several surveys are conducted independently to estimate the same percentages and total numbers in different classes of a population. Consider a population proportion P. Let p_i, $i = 1, 2, ..., k$, denote its estimators from k independent samples of sizes n_i. The variances $V(p_i)$ of these estimators and their sample estimates $v(p_i)$ are obtained from (4.7) and (4.8) by replacing n by n_i.

An unbiased estimator for P is given by the weighted average $p = \Sigma(n_i/n)p_i$, where $n = \Sigma n_i$ is the overall sample size. The variance of p is $\Sigma(n_i/n)^2 V(p_i)$, which is smaller than the variances of the individual p_i, and it is estimated from $\Sigma(n_i/n)^2 v(p_i)$. If the population size and the sample sizes are large, the variance of p can be estimated from $\Sigma(n_i/n)^2 p_i(1 - p_i)/n_i$.

Example 4.5. Combining independent estimates: Estimates of the percentages for the three candidates can be obtained by combining the five surveys in Table 4.1 assuming that they are independent. The total sample size is $n = 15,882$.

For Mr. Clinton, since each p_i is equal to 0.44, the overall estimate is also $p = 0.44$, and $v(p) = 0.44(0.56)/15,882$ and S.E.$(p) = 0.0039$. The 95% confidence intervals for the proportion are $0.44 - 1.96(0.0039) = 0.43$ and $0.44 + 1.96(0.0039) = 0.45$. Thus, the overall estimate is 44% and the confidence limits are $(43, 45)$%.

Similarly, for Mr. Bush, the overall estimate is 37% and the confidence limits are $(36, 38)$%. For Mr. Perot, the overall estimate is 15% and the confidence limits are $(14, 16)$%. As can be seen for each of the three candidates, the standard errors of the combined estimates are smaller than the standard errors found in Example 4.4 from any one of the surveys, and the confidence widths are smaller.

4.9   Covariances of proportions

As will be seen below, for comparing the proportions in the classes, variances as well as covariances of their estimates will be needed. To obtain expressions for the covariances, note that

$$V(p_1 + p_2 + \cdots + p_k) = V(p_1) + V(p_2) + \cdots + V(p_k)$$
$$+ 2\,\text{Cov}(p_1, p_2) + 2\,\text{Cov}(p_1, p_3) + \cdots.$$
$$(4.22)$$

Since $(p_1 + p_2 + \cdots + p_k) = 1$, the variance on the left-hand side vanishes. The variances $V(p_i)$ are given by $(N - n)P_i(1 - P_i)/(N - 1)n$. Substituting these expressions in (4.22), for $(i \neq j)$, $(i, j) = 1, 2, ..., k$,

$$\text{Cov}(p_i, p_j) = -\frac{N - n}{N - 1}\frac{P_i P_j}{n} \qquad (4.23)$$

Similarly, the estimator of this covariance is

$$\hat{\text{Cov}}(p_i, p_j) = -\frac{N - n}{N}\frac{p_i p_j}{n - 1} \qquad (4.24)$$

The expressions in (4.23) and (4.24) can also be obtained directly from the probability distribution in (4.20).

4.10　Difference between proportions

Comparisons of the proportions and the numbers of units in the different classes are frequently of interest. An unbiased estimator of $(P_i - P_j)$ is $(p_i - p_j)$. Its variance and estimator of variance are

$$V(p_i - p_j) = V(p_i) + V(p_j) - 2\text{Cov}(p_i, p_j)$$

$$= \frac{N - n}{(N - 1)n}[P_i(1 - P_i) + P_j(1 - P_j) + 2P_iP_j] \quad (4.25)$$

and

$$v(p_i - p_j) = \frac{N - n}{N(n - 1)}[p_i(1 - p_i) + p_j(1 - p_j) + 2p_ip_j]. \quad (4.26)$$

An estimate for the difference $(C_i - C_j)$ is $N(p_i - p_j)$. The sample S.E. of this estimate is obtained from $N\sqrt{v(p_i - p_j)}$. As can be seen from these expressions, the standard errors of the estimates of $(P_i - P_j)$ and $(C_i - C_j)$ are larger than the standard errors of the individual proportions or totals.

Example 4.6. Educational standards: As reported in Newsweek magazine (May, 1988), in a survey conducted by the Gallop organization in 1987, in a sample of 542 college students, 62% thought that they were receiving a "better" education than their parents' generation, 7% said "worse," and 25% thought that it was about the same. The remaining 6% were the "Don't knows."

An estimate of the difference of the proportions in the first and third categories is $(0.62 - 0.25) = 0.37$ or 37%. With the assumption of a simple random sample for the above survey and a large size for the population, from (4.26), the variance of this estimate is $[0.62(0.38) + 0.25(0.75) + 2(0.62)(0.25)]/541 = 13.55 \times 10^{-4}$, and the S.E. is 0.0368 or about 3.7%. The 95% confidence limits for the difference are given by $0.37 \pm 1.96(0.0368)$, that is, 30 and 44%.

4.11　Sample sizes for more than two classes

When there are more than two classes, the sample size required to estimate the proportions and totals can be found from the type of specifications described in Section 4.5. Similarly, sample size to estimate the difference $(P_i - P_j)$ of two proportions can be found by replacing $V(p_i)$ by $V(P_i - P_j)$ in the corresponding expressions. If no information on the proportions is available, as before the maximum sample size is obtained by replacing each proportion by 0.5.

Example 4.7. Sample sizes for two or more classes: Consider more than two classes, as in the case of the political and public polls described in Section 4.6, and the problem of estimating their proportions. If the error of estimation for estimating any of the proportions should not exceed 0.05 except for a 5% chance, the sample size is obtained from (4.16). If it is known, for example, that the maximum of the percentages is 45, the required sample size is given by $(1.96)^2(0.45)(0.55)/(0.05)^2 = 380.32$ or 381 approximately. If no information is available on the percentages, as in Example 4.3, a sample size of at least 384 would be needed.

To estimate the difference $(P_i - P_j)$ of two proportions with an error not exceeding 0.05 except for a 5% chance, the required sample size is obtained by equating $e/S.E.(p_i - p_j)$ to 1.96. With 0.45 for each proportion, from (4.25), the variance of $(p_i - p_j)$ for large N equals $[(0.45)(0.55) + (0.45)(0.55) + 2(0.45)(0.55)] = 0.9/n$. Thus, the required sample size is equal to $(1.96)^2(0.9)/(0.05)^2 = 1383$. With 0.5 for each proportion, the sample size would be $(1.96)^2/(0.05)^2 = 1537$ approximately. In either case, sample sizes required for estimating the difference are larger than the sizes for the individual proportions. This result is expected since the S.E. of the difference of the sample proportions is larger than the S.E. of an individual sample proportion.

4.12 Estimating population size

Throughout the previous sections, the population size N was assumed to be known. To estimate the sizes of animal, fish, and other types of moving populations, the **capture–recapture** method can be used. In this procedure, a sample of n_1 units observed from the population is tagged or marked and released. The number of units c appearing again in an independent sample of n_2 units is noted. The probability distribution of these common units is

$$P(c) = \binom{n_1}{c}\binom{N - n_1}{n_2 - c} \bigg/ \binom{N}{n_2} = \binom{n_2}{c}\binom{N - n_2}{n_1 - c} \bigg/ \binom{N}{n_1}, \quad (4.27)$$

where $0 \le c \le \min(n_1, n_2, N - n_1 - n_2)$.

The proportion of the units tagged in the first sample is n_1/N. An unbiased estimator of this proportion is c/n_2. Hence, $\hat{N} = n_1 n_2/c$ can be considered to be an estimator of N. Another motivation for $\hat{N}$ arises from noting that $E(c) = n_1 n_2/N$. It is, however, an overestimate of N since $E(1/c) \geq 1/E(c)$.

For large N, (4.27) approximately becomes the binomial distribution

$$P(c) = \binom{n_2}{c} P^c Q^{n_2 - c} \tag{4.28}$$

where $P = (n_1/N)$. Now, (c/n_2) is an unbiased estimator of this proportion. From this result, again $n_1 n_2/c$ can be considered an estimator for N.

Exercises

4.1.   Enrollments for higher education beyond high school for the 50 states in the U.S. and the District of Columbia (DC) are available in the *Statistical Abstracts of the United States*; figures for 1990 are available in the 1992 publication. In a random sample of 10 from 49 states and DC, excluding the two largest states, New York and California, enrollments in 1990 exceeding 100,000 were observed for public institutions in six states and for private institutions in one state. (a) For each type of institution, find the estimate and its S.E. for the number of states in which enrollments exceeded 100,000. (b) Estimate the difference of the above numbers for the two types and find the S.E. of the estimate.

4.2.   If no information on P is available, find the error of estimation e in (4.15) for (a) $n = 1000$ and (b) $n = 2000$. Consider $\alpha = 0.05$ for both cases.

4.3.   The sampling errors for the first two and last two public polls in Table 4.2 were reported to be 4 and 4.5%, respectively. Explain how these results are obtained, noting that the estimate in each case should not differ from the population percentage by the specified amount of error except for a 5% chance.

4.4.   From the sample data in Table 4.2, for the change from September to October in the percentage favoring the health care program, find (a) the estimate, (b) its S.E., and (c) the 95% confidence limits.

4.5. From the results in Table 4.2 for September, for the percentage not in favor, that is, opposing and not sure, find (a) the estimate, (b) its S.E., and (c) the 95% confidence limits. (d) Test the hypothesis at $\alpha = 0.05$ that the percentage favoring is not more than 10% from the percentage not favoring.

4.6. From the figures in Table 4.3, for the participation of both the Democrats and Republicans together, find the estimate, its S.E., and the 95% confidence limits for the (a) percentage of men, (b) percentage of women, and (c) difference in the percentages for men and women.

4.7. From the figures in Table 4.3, for the percentage of the participants in the primaries having at least some college education, find the estimate, its S.E., and the 95% confidence limits for (a) Democrats, (b) Republicans, and (c) both Democrats and Republicans together. (d) Describe how one can find these estimates, standard errors, and confidence limits from the percentages of those who do not have any college education.

4.8. For the survey on 1038 teenagers described in Section 1.8, 75 and 81% of the boys and girls, respectively, emphasized the importance of good grades. For 53 and 41% of the boys and girls, it was important to have a lot of friends. (a) For both good grades and a lot of friends, find the 95% confidence limits for the difference in the percentages for boys and girls. (b) For good grades and for having a lot of friends, test the hypothesis at the 5% level that there is no significant difference between boys and girls. Assume that the responses were obtained from equal number, 519, of boys and girls.

4.9. One university has 1000 students in each of the four classes. The percentages of the Freshman–Sophomore, Junior, and Senior classes expressing interest in professional training after graduation were guessed to be 20, 50, and 80%, respectively. (a) For each of these three groups, find the sample sizes required to estimate the percentage if the estimate should not differ from the actual value by not more than 20% of the actual value except for a 5% chance, and present the reason for the differences in the sample sizes. (b) Find the sample sizes needed for each of the three groups for estimating the above percentage if the error of estimation should not exceed 10% except for a 5% chance, and present the reason for the differences in the sample sizes.

4.10.   Among the 637 families in a village in New York State, 241 have children currently attending school. In the remaining 396 families, either the children have completed schooling or they have no children. In a straw poll, 14 families from the first group favored the proposal to increase the school budget, two had no opinion, but nine opposed. The corresponding figures for the second group are 9, 3, and 12. For the proportion and number of families in the village who have no opinion, find (a) the estimate, (b) its S.E., and (c) the 95% confidence limits. For the difference in the proportions and numbers of the families in the two groups favoring the proposal, find (a) the estimate, (b) its S.E., and (c) the 95% confidence limits.

4.11.   With the preliminary estimates obtained from the data in Exercise 4.10, find the sample sizes needed to estimate the percentage of the families favoring the proposal (a) with an error less than 10% except for a 1 in 20 chance, (b) with the error not exceeding 1/10 of the actual percentage except for a 5% chance, and (c) with the confidence width for a 90% confidence probability not exceeding 10%.

4.12.   Using the preliminary estimates from the data in Exercise 4.10, find the sample sizes required to estimate the total number of families favoring the proposal (a) if the error of estimation should not exceed 10% of the actual number except for a 5% chance and (b) if the coefficient of variation of the estimate should not be more than 5%.

4.13.   At the end of a cultural conference at a university, 110 faculty, 250 students, and 200 other participants returned the questionnaires on their opinions regarding the conference. Among these three groups, 80, 120, and 100 were in favor of continuing the conference every year. Denote the actual proportions in the three groups favoring the continuation by P_1, P_2, and P_3. Using the preliminary estimates from these responses, find the overall sample sizes needed (a) to estimate $(P_1 - P_2)$ with an S.E. not exceeding 7%, and (b) to estimate the difference between the proportions for any two groups with an S.E. not exceeding 7%. The university community consists of 500, 6000, and 2000 persons in the above three categories.

4.14. In the survey described in Exercise 3.9, 12 of the 15 respondents suggested that the campus should have more tennis courts. Thus, for this characteristic, the sample estimate of the proportion for the respondents is $p_1 = 12/15 = 0.8$. For the population proportion, three estimates may be considered: (1) p_1, (2) substitute p_1 for each of the nonrespondents, and (3) substitute zero for each of the nonrespondents. Find expressions for the biases, variances, and MSEs of these three procedures.

4.15. From (4.20), show that $E(c_1 c_2) = n(n - 1)C_1 C_2/N(N - 1)$. Using this result, (a) derive the covariance in (4.23) and (b) show that the expression in (4.24) is unbiased for this covariance.

Appendix A4

Unbiasedness of the sample proportion and variance

With the probability in (4.9),

$$E(c) = \sum_c c \, \Pr(c, n - c) = nP$$

and hence $P = c/n$ is unbiased for P.
Similarly,

$$E[c(c - 1)] = \frac{n(n - 1)}{N(N - 1)} C(C - 1),$$

which can be expressed as

$$E[np(np - 1)] = \frac{n(n - 1)}{N(N - 1)} NP(NP - 1),$$

From this expression,

$$E(p^2) = \left[\frac{N-n}{N-1} + \frac{N(n-1)P}{N-1}\right]\frac{P}{n}.$$

Finally writing $V(p)$ as $E(p^2) - P^2$, the expression for the variance becomes the same as (4.7).

From this result, an unbiased estimator for P^2 is given by $[(N-1)np^2 - (N-n)p]/N(n-1)$.

Since $PQ = P - P^2$, its unbiased estimator from the above results is given by $p - [(N-1)np^2 - (N-n)p]/N(n-1)$, that is,

$$\hat{PQ} = \frac{(N-1)n}{N(n-1)}pq.$$

By utilizing this result, (4.8) becomes the unbiased estimator of the variance in (4.7).

Binomial approximation

The hypergeometric distribution in (4.9) can be expressed as

$$\Pr(c, n-c) = \frac{n!}{c!(n-c)!}\frac{C!(N-C)!(N-n)!}{(C-c)!(N-C-n+c)!N!}$$

If N is large, this expression becomes

$$\Pr(c, n-c) = \frac{n!}{c!(n-c)!}P^c Q^{n-c},$$

which is the **binomial** distribution.

This is the probability for c *successes* in n independent trials, with the probability of success P remaining the same at every trial—the **Bernoulli** trials. For this distribution, $p = c/n$ is unbiased for P. Its variance and estimator of variance are $V(p) = PQ/n$ and $v(p) = pq/n$.

CHAPTER 5

Stratification

5.1   Introduction

Each layer of a rock is a **stratum**, and the layers are the **strata**. There is usually a good deal of uniformity in the mineral contents and the composition of each of these layers. The Himalayas and Rockys are ranges of mountains. Each mountain is a stratum, and the ranges are the strata. Composition and fertility of the soil along the banks of a river vary as it gallops from its birth place to the ocean. The entire stretch can be divided into strata based on this variation. These types of analogies are used in survey sampling to describe the subdivisions or partitions of a population. The observations of the units of a stratum are closer to each other than to the units of another stratum.

As was seen in Section 1.8, for some types of income and expenditure surveys on households in urban areas, states, provinces, counties, and districts may be considered as the strata. For business surveys on employee size, production, and sales, stratification is usually based on industrial classifications. For agricultural surveys in rural areas, villages and geographical regions compose the strata. In general, stratification of population units depends on the purpose of the survey.

The major advantages of stratification are (1) estimates for each stratum can be obtained separately, (2) differences among the strata can be evaluated, (3) the total, mean, and proportion of the entire population can be estimated with high precision by suitably weighting the estimates obtained from each stratum, and (4) there are frequently savings in time and cost needed for sampling the units. In addition, it is usually convenient to sample separately from the strata rather than from the entire population, especially if the population is large.

The following sections examine the estimation of the mean, total, and proportion of each stratum as well as the entire population, determination of the sample sizes for the strata for specified requirements, and other topics of interest in stratified random sampling.

5.2   Notation

For the sake of illustration, Table 5.1 presents the rice production in 1985 in 43 countries. These figures for 65 of the 191 countries in the world appear in the *Statistical Abstracts of the United States, 1990.* The amounts for the three largest producing countries, China, India, and Indonesia, appear as 171.5, 91.5, and 38.7 million metric tons. Two of the countries produce only 3000 metric tons. Productions for 17 countries are in the NA (not available) category. Based on their production levels, the remaining 43 countries have been divided into two strata (groups) of sizes 32 and 11.

The total, mean, variance, and standard deviation for the population of the $N = 43$ productions in Table 5.1 are $Y = 157.27$, $\bar{Y} = 3.66$, $S^2 = 31.68$, and $S = 5.63$.

When a population consists of G strata of sizes N_g, the observations of the strata can be represented by y_{gi}, $g = 1, 2,...,G$ and $i = 1, 2,...$, N_g. In Table 5.1, the population of $N = 43$ productions is divided into $G = 2$ strata of sizes $N_1 = 32$ and $N_2 = 11$. The observations for the first stratum are $y_{11} = 0.04$, $y_{12} = 0.04$, $y_{13} = 0.06,...,y_{1,32} = 2.80$. Similarly, for the second stratum, $y_{21} = 4.50$, $y_{22} = 5.60$, $y_{23} = 6.17,...$, $y_{2,11} = 21.90$.

Table 5.1. Rice production (in million metric tons).

	First Stratum						Second Stratum
1	0.04	12	0.30	23	1.10	1	4.50
2	0.04	13	0.40	24	1.40	2	5.60
3	0.06	14	0.42	25	1.76	3	6.17
4	0.08	15	0.46	26	1.90	4	7.86
5	0.09	16	0.47	27	1.90	5	8.30
6	0.11	17	0.48	28	2.18	6	9.02
7	0.11	18	0.52	29	2.31	7	14.58
8	0.15	19	0.86	30	2.60	8	15.40
9	0.16	20	0.97	31	2.63	9	15.60
10	0.20	21	0.99	32	2.80	10	19.52
11	0.27	22	1.06			11	21.90
Total					28.82		128.45
Mean					0.90		11.68
Variance					0.79		35.55
S.D.					0.89		5.96

Source: Statistical Abstracts of the United States (1990), Table 1412.

The total and mean of the observations of the gth stratum are given by

$$Y_g = \sum_1^{N_g} y_{gi} \quad \text{and} \quad \bar{Y}_g = Y_g / N_g. \tag{5.1}$$

The variance of this stratum is

$$S_g^2 = \frac{\sum_1^{N_g} (y_{gi} - \bar{Y}_g)^2}{N_g - 1}. \tag{5.2}$$

The expressions in (5.1) and (5.2) are the same as those of the total, mean, and variance presented in Chapter 2, except that the different strata are represented by the subscript g. These summary figures for the two strata appear in Table 5.1.

The total and mean for the entire population are

$$Y = \sum_1^{G} \sum_1^{N_g} y_{gi} = \sum_1^{G} Y_g = \sum_1^{G} N_g \bar{Y}_g \tag{5.3}$$

and

$$\bar{Y} = \frac{Y}{N} = \sum_1^{G} W_g \bar{Y}_g = W_1 \bar{Y}_1 + W_2 \bar{Y}_2 + \cdots + W_G \bar{Y}_L, \tag{5.4}$$

where $W_g = N_g/N$ is the proportion of the population units in the gth stratum. In the right-hand side of (5.3), the first term is the sum of all the observations in the sample, the second term is the sum of the stratum totals, and in the final term the strata means are weighted by their sizes. In (5.4), the population mean is expressed as a weighted average of the means of the strata. As a verification, the population mean of 3.66 for the 43 countries is the same as $(32/43)(0.90) + (11/43)(11.68)$.

The population variance is

$$S^2 = \frac{\sum_1^{G} \sum_1^{N_g} (y_{gi} - \bar{Y})^2}{N - 1}. \tag{5.5}$$

The numerator of this expression, $(N - 1)S^2$, is the sum of squares of the deviations from the population mean of all the observations in all the strata, and hence it is known as the **total SS**.

As shown in Appendix A5, (5.5) can also be expressed as

$$S^2 = \frac{\sum_1^G (N_g - 1)S_g^2 + \sum_1^G N_g(\bar{Y}_g - \bar{Y})^2}{N - 1}.$$

(5.6)

If N_g and N are large enough that they differ little from $(N_g - 1)$ and $(N - 1)$, this variance can be approximately expressed as $\Sigma W_g S_g^2 + \Sigma W_g(\bar{Y}_g - \bar{Y})^2$. The first term in the numerator of (5.6) is the **within SS** (sum of squares), which is the *pooled* sum of squares of the squared deviations of the observations of the strata from their means. The second term of (5.6) which expresses the variation among the strata means is the **between SS**. Some of the S_g^2 may be larger than S^2, but it is larger than the weighted average of the S_g^2.

5.3   Estimation for a single stratum

Samples of sizes n_g are selected from the N_g units of the strata, randomly without replacement and independently from the strata. The general description of the strata and the samples are presented in Table 5.2.

The mean and variance of a sample of n_g units selected from the gth stratum are

$$\bar{y}_g = \sum_1^{n_g} y_{gi} / n_g$$

(5.7)

and

$$s_g^2 = \frac{\sum_1^{n_g} (y_{gi} - \bar{y}_g)^2}{n_g - 1}.$$

(5.8)

Table 5.2. Strata and samples.

	1	2	g	G
Strata				
Sizes	N_1	N_2	N_g	N_G
Means	$\bar{Y}_1$	$\bar{Y}_2$	$\bar{Y}_g$	$\bar{Y}_G$
Variances	S_1^2	S_2^2	S_g^2	S_G^2
Samples				
Sizes	n_1	n_2	n_g	n_G
Means	$\bar{y}_1$	$\bar{y}_2$	$\bar{y}_g$	$\bar{y}_G$
Variances	s_1^2	s_2^2	s_g^2	s_G^2

As seen in Chapter 2, this mean and variance are unbiased for $\overline{Y}_g$ and S_g^2, respectively. The variance of $\bar{y}_g$ is

$$V(\bar{y}_g) = \frac{N_g - n_g}{N_g n_g} S_g^2 = \frac{(1 - f_g)}{n_g} S_g^2, \tag{5.9}$$

where $f_g = n_g/N_g$ is the sampling fraction in the gth stratum. The estimator,

$$v(\bar{y}_g) = \frac{N_g - n_g}{N_g n_g} s_g^2 = \frac{(1 - f_g)}{n} s_g^2 \tag{5.10}$$

is unbiased for the above variance.

The expressions in (5.7) through (5.10) with the additional subscript g representing the stratum are the same as those presented in Chapter 2 for the mean of a simple random sample.

For the total Y_g of the gth stratum, an unbiased estimator is given by $N_g \bar{y}_g$. Its variance and estimator of variance are obtained by multiplying (5.9) and (5.10) by N_g^2.

Example 5.1. Rice production: For the sake of illustration, consider a sample of size $n = 10$ from the $N = 43$ countries in Table 5.1, and distribute this sample size proportional to the sizes of the two strata. Thus, the sample sizes are $n_1 = (0.74)n$ and $n_2 = (0.26)n$, which are approximately equal to 7 and 3. For these sample sizes, from (5.9), $V(\bar{y}_1) = [(32 - 7)/32] (0.09)/7 = 0.0882$, S.E.$(\bar{y}_1) = 0.297$, $V(\bar{y}_2) = [(11 - 3)/11](35.55)/3 = 8.62$, and S.E. $(\bar{y}_2) = 2.94$.

To examine the estimates from the samples, samples of the above sizes were selected randomly without replacement through the Random Number Table in Appendix T1, independently from the two groups. Through this procedure, countries (4, 6, 9, 12, 17, 22, 31) and (1, 6, 10) appear in the samples of the two groups, respectively. The means and variances of these samples are $\bar{y}_1 = 0.69$, $s_1^2 = 0.85$, $\bar{y}_2 = 11.01$, and $s_2^2 = 59.38$.

Now, from (5.10), $v(\bar{y}_1) = [(32 - 7)/32](0.85/7) = 0.095$, and hence, S.E.$(\bar{y}_1) = 0.31$. Similarly, $v(\bar{y}_2) = [(11 - 3)/11](59.38/3) = 14.395$ and S.E.$(\bar{y}_2) = 3.79$.

5.4   Estimation of the population mean and total

With stratification, an estimator for the population mean $\bar{Y}$ is given by

$$\hat{\bar{Y}}_{st} = \sum_1^G W_g \bar{y}_g = W_1 \bar{y}_1 + W_2 \bar{y}_2 + \cdots + W_G \bar{y}_G, \qquad (5.11)$$

where the subscript st denotes stratification. This estimator is obtained by substituting the sample means for the stratum means in (5.4). Note that the sample means are multiplied by the strata weights. Since $E(\bar{y}_g) = \bar{Y}_g$, this weighted mean is unbiased for $\bar{Y}$. Since the samples are selected independently from the strata, as shown in Appendix A5, the covariances between the sample means vanish and the variance of $\hat{Y}_{st}$ becomes

$$V(\hat{\bar{Y}}_{st}) = \sum_1^G W_g^2 V(\bar{y}_g) = W_1^2 V(\bar{y}_1) + W_2^2 V(\bar{y}_2) + \cdots + W_G^2 V(\bar{y}_G).$$
$$(5.12)$$

Substituting the variances of the strata $V(\bar{y}_g)$ from (5.9), this variance can be expressed as

$$V(\hat{\bar{Y}}_{st}) = \sum_1^G W_g^2 \frac{(1 - f_g)}{n_g} S_g^2 = \sum \frac{W_g^2 S_g^2}{n_g} - \frac{\sum W_g S_g^2}{N}. \qquad (5.13)$$

An unbiased estimator of this variance is obtained by replacing $V(\bar{y}_g)$ in (5.12) by $v(\bar{y}_g)$ in (5.10), that is, replacing S_g^2 in (5.13) by s_g^2. Thus,

$$v(\hat{\bar{Y}}_{st}) = \sum_1^G W_g^2 v(\bar{y}_g) = \sum_1^G W_g^2 \frac{(1 - f_g)}{n_g} s_g^2. \qquad (5.14)$$

An unbiased estimator of the population total Y is $N \hat{\bar{Y}}_{st}$. Notice that this estimator is the same as $N_1 \bar{y}_1 + N_2 \bar{y}_2 + \cdots + N_G \bar{y}_G$. Thus, the population total is estimated by adding the estimates for the totals of all the strata. Its variance and estimator of variance are obtained by multiplying (5.13) and (5.14) by N^2.

Example 5.2. Rice production in the world: With the sample means in Example 5.1, the estimate for the mean of the 43 countries is $\hat{Y}_{st} =$ $(32/43)(0.69) + (11/43)(11.01) = 3.33$. From the first term on the right-hand side of (5.13), $V(\hat{Y}_{st}) = (32/43)^2(0.088) + (11/43)^2(8.62) = 0.613$, and S.E.$(\hat{Y}_{st}) = 0.78$. For the estimate of the variance, from (5.14), $v(\hat{Y}_{st}) = (32/43)^2(0.095) + (11/43)^2(14.395) = 0.9946$, and hence S.E. $(\hat{Y}_{st}) = 0.9973$.

Now, an estimate for the total rice production for the 43 countries is $\hat{Y}_{st} = 43(3.33) = 143.19$ and S.E.$(\hat{Y}_{st}) = 43(0.9973) = 42.88$. From the sample, the estimate for S.E.$(\hat{Y}_{st})$ is $43(0.78) = 33.54$.

By adding the 301.7 million pounds of the three largest producing countries to this estimate, the estimate for the world rice production is close to 445 million metric tons a year. The actual production, as appears in *Statistical Abstracts*, is 466 million metric tons.

Note that without stratification, for selecting a simple random sample of 10 from the 43 units, from (2.12), $V(\bar{y}) = [(43 - 10)/43]$ $(31.68/10) = 2.431$. Thus, the precision of $\bar{Y}_{st}$ relative to $\bar{y}$ is $2.431/0.613 = 3.97$. This result shows that there is a 297% gain in precision for stratification.

As noted in Section 3.6, Cochran (1977, pp. 215–216) shows that if there is a linear trend in the population units, systematic sampling results in smaller variance for the mean than for simple random sampling. He also shows that for this case, the variance of $\hat{Y}_{st}$ is smaller than the means of both simple random and systematic sampling. Further comparisons of these three procedures appear in Cochran (1946).

5.5 Confidence limits

As in Section 2.9, $(1 - \alpha)\%$ confidence limits for $\bar{Y}_g$ are obtained from $\bar{y}_g \pm Z.$ S.E.$(\bar{y}_g)$. Similarly, one can obtain the confidence limits for $\bar{Y}$ from $\bar{Y}_{st} \pm Z.$ S.E.$(\bar{Y}_{st})$. Multiply these limits by N to obtain the limits for the population total Y.

From the results of Exercise 5.2, the 95% confidence limits for average rice production for the 43 countries are given by $3.33 \pm 1.96(0.9973)$, that is, $(1.38, 5.29)$.

When S_g^2 are estimated, the limits may be obtained from $\bar{y}_g \pm t$ S.E.$(\bar{y}_g)$. Approximation for the d.f. of the t-distribution needed to find these limits is presented, for example, by Cochran (1977, p. 96).

5.6   Proportions and totals

The results of Chapter 4 and Sections 5.3 and 5.4 can be combined to estimate for each stratum and the population the proportions and total numbers of units having a specific qualitative characteristic. All the expressions in this chapter for a single stratum are easily obtained from Chapter 4 with the subscript g for the gth stratum

Let C_g and $P_g = C_g/N_g$, $g = 1,...,G$, denote the number and proportion of units in the gth stratum having the characteristic of interest. If c_g units in a random sample of size n_g are observed to have the characteristic, the sample proportion $p_g = c_g/n_g$ is unbiased for P_g. Its variance and estimator of variance are

$$V(p_g) = \frac{N_g - n_g}{N_g - 1} \frac{P_g Q_g}{n_g} \qquad (5.15)$$

and

$$v(p_g) = \frac{N_g - n_g}{N_g} \frac{p_g q_g}{n_g - 1}, \qquad (5.16)$$

where $Q_g = 1 - P_g$ and $q_g = 1 - p_g$. These expressions are obtained from (4.7) and (4.8) with the subscript g for the stratum. An unbiased estimator for C_g is given by $\hat{C}_g = N_g p_g$. For the variance of $\hat{C}_g$ and its estimator, multiply (5.15) and (5.16) by N_g^2.

Example 5.3. Energy consumption: For the sake of illustration, energy production and consumption in 1996 for the 25 countries in the world with the highest per capita consumption is presented in Table T9 in the Appendix. For 15 of these countries, that is, 60%, per capita consumption exceeds 200 million BTUs.

To study the production and consumption of energy, one can divide the 25 countries into two strata with the first stratum consisting of countries producing more than 200,000 barrels of petroleum a day, and the second stratum consisting of the rest. From this division, the 10 countries numbered (1, 2, 6, 8, 14, 17, 18, 19, 23, 24) are included in the first stratum, and the remaining 15 in the second stratum.

Among the $N_1 = 10$ countries, for $C_1 = 7$ countries energy consumption exceeds 200 million BTUs. Thus, $P_1 = 7/10$ for this characteristic. Corresponding figures for the second stratum are $N_2 = 15$, $C_2 = 8$, and $P_2 = 8/15$. If these percentages are not known, one can estimate them from samples selected from the two strata. For example, for a sample of $n_1 = 3$

units from the first stratum, from (5.15), $V(p_1) = [(10 - 3)/(9 \times 3)](7/10)$ $(3/10) = 0.0544$. Similarly, for a sample of size $n_2 = 5$ from the second stratum, $V(p_2) = [(15 - 5)/(14 \times 5)](8/15)(7/15) = 0.0356$.

5.7 Population total and proportion

The total and proportion of the population having a specific characteristic, such as the number of countries with per capita energy consumption exceeding 200 million BTUs, are

$$C = \sum_1^G C_g = C_1 + C_2 + \cdots + C_G$$
$$= \sum N_g P_g = N_1 P_1 + N_2 P_2 + \cdots + N_G P_G \qquad (5.17)$$

and

$$P = C/N = \sum W_g P_g. \qquad (5.18)$$

An unbiased estimator for P is given by

$$\hat{P}_{st} = \sum_1^G W_g p_g \qquad (5.19)$$

with variance

$$V(\hat{P}_{st}) = \sum_1^G W_g^2 V(p_g) = \sum W_g^2 \frac{N_g - n_g}{N_g - 1} \frac{P_g Q_g}{n_g} \qquad (5.20)$$

An unbiased estimator for this variance is given by

$$V(\hat{P}_{st}) = \sum W_g^2 (1 - f_g) \frac{p_g q_g}{n_g - 1}. \qquad (5.21)$$

An unbiased estimator for the total number of units having the characteristic is $\hat{C}_{st} = N\hat{P}_{st}$, which has variance $N^2 V(\hat{P}_{st})$ and estimator of variance $N^2 v(\hat{P}_{st})$. The S.E. of $\hat{P}_{st}$ from the sample is given by $N\sqrt{v(\hat{P}_{st})}$. For large sizes of the population and sample, confidence limits for P can be obtained from $\hat{P}_{st} \pm Z$ S.E.$(\hat{P}_{st})$. For C, multiply these limits by N.

Example 5.4. Energy consumption: For the two strata in Example 5.3, $W_1 = 10/25 = 0.4$ and $W_2 = 15/25 = 0.6$. With samples selected from the two strata, the proportion of the countries with per capita consumption exceeding 200 million BTUs is estimated from $\hat{P}_{st} = W_1 p_1 + W_2 p_2$. For samples of sizes $n_1 = 3$ and $n_2 = 5$ from the strata, we have found in Example 5.3 that $V(p_1) = 0.0544$ and $V(p_2) = 0.0356$. From these figures, $V(\hat{P}_{st}) = (0.4)^2(0.0544) + (0.6)^2(0.0356) = 0.0215$, and hence S.E.$(\hat{P}_{st}) = 0.1466$. For the estimation of the total number of countries with the above characteristic, $V(\hat{C}_{st}) = 25^2(0.0215) = 13.44$. From this variance or directly, S.E.$(\hat{C}_{st}) = 25(0.1466) = 3.67$.

5.8   Proportional and equal division of the sample

If the resources are available for a total sample of n units, they are distributed among the strata. One obvious choice is to divide this sample size proportionate to the stratum sizes; that is, select $n_g = n(N_g/N) = nW_g$ units from the gth stratum. As an alternative, if it is convenient and cost-effective, one may select equal number, $n_g = n/G$, units from the strata. An examination of the precision for these two types of determining the sample sizes follows.

Proportional allocation

For this type of dividing the sample, since $n_g/n = N_g/N = W_g$, the sampling fraction $f_g = n_g/N_g$ is the same for all the strata and is equal to the overall sampling fraction $f = n/N$. The estimator $\bar{Y}_{st}$ in (5.11) now becomes **self-weighting** and can be expressed as

$$\hat{\bar{Y}}_{st} = \frac{\sum^{G} n_g \bar{y}_g}{n} = \frac{\sum^{G} \sum^{n_g} y_{gi}}{n}. \tag{5.22}$$

This estimate is the average of all the n observations. Notice, however, that in general $\hat{\bar{Y}}_{st}$ in (5.11) is the weighted average of the means $\bar{y}_g$ with the strata weights W_g and it is unbiased for $\bar{Y}$ for any type of allocation of the sample.

For proportional allocation, replacing n_g by $n(N_g/N)$, the variance in (5.13) becomes

$$V_P = \frac{(1-f)}{n} \sum_1^G W_g S_g^2. \tag{5.23}$$

Notice that $\sum_1^G W_g S_g^2$ is the weighted average of the stratum variances. For the rice production in Example 5.2, from (5.23), $V_P = [(43 - 10)/430][(32/43)(0.79) + (11/43)(35.55)] = 0.743$. However, $V(\bar{Y}_{st})$ was found to be 0.613 since the sample sizes were rounded off to 7 and 3 for the strata.

From (2.12), the variance of the sample mean is $V_S = (1-f)S^2/n$. As noted in Section 5.2, S^2 can be approximately expressed as $\sum W_g S_g^2 + \sum W_g(\bar{Y}_g - \bar{Y})^2$. This approximation is valid when $N_g - 1$ and $N - 1$ do not differ much from N_g and N, respectively. Substituting this expression in V_S, from (5.23), $V_S - V_P = [(1-f)/n]\sum W_g(\bar{Y}_g - \bar{Y})^2$, which is non-negative. Thus, proportional allocation results in smaller variance than simple random sampling if the means $\bar{Y}_g$ differ much.

From (2.12) and (5.23) the precision of proportional allocation relative to simple random sampling is

$$\frac{V_S}{V_P} = \frac{S^2}{\sum W_g S_g^2}. \tag{5.24}$$

Since $S^2 > \sum W_g S_g^2$ from (5.6), this relative precision is at least 100%. Notice, however, that it does not depend on the sample size.

Equal distribution

As an illustration of assigning equal sample sizes to all the strata, consider a population with five strata and the resources available for a total sample of 100 units. In this case, five interviewers can collect information on 20 sample units of one stratum each or the survey can be completed in 5 days by collecting the information on 20 sample units of a stratum per day. For this type of sampling, the variance of $\hat{\bar{Y}}_{st}$ and its estimator are obtained from (5.13) and (5.14) by substituting

n/G for n_g. The following example compares this approach with proportional allocation.

> **Example 5.5.** Proportional and equal allocation: For the rice production in Example 5.2, with $n_1 = 7$ and $n_2 = 3$ for proportional allocation, $V_P = 0.613$. For equal allocation with $n_1 = n_2 = 5$, from (5.13), $V(\bar{y}_1) = [(32 - 5)/32](0.79)/5 = 0.1333$ and $V(\bar{y}_2) = [(11 - 5)/11](35.55)/5 = 3.88$. Now, from (5.13), $V(\hat{Y}_{st}) = (32/43)^2(0.1333) + (11/43)^2(3.88) = 0.328$. By denoting this variance for equal allocation by V_E, $V_E/V_P = 0.328/0.613 = 0.54$. Thus, equal allocation reduces the variance by 46% from proportional allocation.
>
> For the rice production in Example 5.2, as has been seen, $N_1 = 32$, $N_2 = 11$, $S_1 = 0.89$, and $S_2 = 5.96$. Proportional allocation considered more sample units for the larger first stratum. Equal allocation resulted in increasing the sample size by two units for the second stratum, which has larger variance. For the division of the sample size for the strata, the following procedure takes into account both the sizes and the variances of the strata.

5.9   Neyman allocation

For a given n, Neyman (1934) suggested finding n_g by minimizing the variance in (5.13). As shown in Appendix A5, from this optimization,

$$\frac{n_g}{n} = \frac{N_g S_g}{\sum N_g S_g} = \frac{W_g S_g}{\sum W_g S_g}. \tag{5.25}$$

Note that it is enough to know the relative values of S_g for determining the sample sizes for the strata. This allocation suggests that variance for $\hat{Y}_{st}$ is reduced when strata with larger sizes N_g and larger variances S_g^2 receive larger sample sizes n_g. With this method, the minimum of the variance in (5.13) becomes

$$V_N = \frac{\left(\sum W_g S_g\right)^2}{n} - \frac{\left(\sum W_g S_g^2\right)}{N}. \tag{5.26}$$

The sample size n_g obtained from (5.25) for one or more strata can be as small as unity in some instances if N_g or S_g are relatively small.

For such strata, $V(\bar{y}_g)$ will be zero. On the other hand, since (5.25) is obtained without the constraint $n_g \leq N_g$, it may result in n_g larger than N_g for some of the strata. In this case, all or a predetermined number of units are sampled from these strata, and the remaining sample size is distributed to the rest of the strata.

5.10   Gains from Neyman allocation

From (5.23) and (5.26),

$$V_P - V_N = \left[\sum W_g S_g^2 - \left(\sum W_g S_g \right)^2 \right]/n$$
$$= \sum W_g \left(S_g - \sum W_g S_g \right)^2 / n. \qquad (5.27)$$

The right-hand side expresses the differences among the standard deviations S_g of the strata. Notice from (5.27) that the variance for Neyman allocation becomes smaller than that for proportional allocation as the differences among the S_g increase. The precision of this allocation relative to proportional allocation is

$$\frac{V_P}{V_N} = \frac{N - n}{N \left(\sum W_g S_g \right)^2 / \sum (W_g S_g^2) - n}. \qquad (5.28)$$

Since $\sum W_g S_g^2 > (\sum W_g S_g)^2$, this ratio is larger than unity, and it also increases with n.

With the approximation for S^2 considered in Section 5.8,

$$V_S - V_N = \sum W_g \left(S_g - \sum W_g S_g \right)^2 / n + (1 - f) \sum W_g (\bar{Y}_g - \bar{Y})^2 / n. \qquad (5.29)$$

The precision of Neyman allocation relative to simple random sampling increases as the differences among the means $\bar{Y}_g$ or the standard deviations S_g increase.

Example 5.6. Neyman allocation: For the rice production strata in Table 5.1, $W_1S_1 = (32/43)(0.89) = 0.66$, $W_2S_2 = (11/43)(5.96) = 1.53$ and $\Sigma W_gS_g = 2.19$. For the Neyman allocation, $n_1/n = 0.66/2.19 = 0.3$ and $n_2/n = 1.53/2.19 = 0.7$. If $n = 10$, the sample sizes for the strata are $n_1 = 3$ and $n_2 = 7$. The second stratum with considerably larger variance than the first receives the larger sample size. Interestingly, in this illustration, the sample sizes are in the opposite direction for proportional allocation. The *equal* allocation is closer to Neyman allocation than proportional allocation.

For Neyman allocation, $V(\bar{y}_1) = [(32 - 3)/32](0.79)/3 = 0.239$ and $V(\bar{y}_2) = [(11 - 7)/11](35.55)/7 = 1.8468$. Now, from (5.13), $V(\bar{Y}_{st}) = (32/43)^2(0.239) + (11/43)^2(1.8468) = 0.253$ and S.E.$(\hat{\bar{Y}}_{st}) = 0.503$. The same values for this variance and S.E. are obtained from (5.26).

As found in Example 5.2, for simple random sampling, the variance of the mean $\bar{y}$ for a sample of ten units is $V_S = 2.431$. Since $V_N/V_S = 0.253/2.431 = 0.104$, Neyman allocation reduces the variance almost by 90%. As found in Example 5.2, for proportional allocation, the variance of $\bar{Y}_{st}$ is $V_P = 0.613$. Now, $V_N/V_P = 0.253/0.613 = 0.413$. Hence, Neyman allocation decreases the variance from proportional allocation by about 58%.

For the $N = 43$ countries in Table 5.1, the precision of Neyman allocation relative to simple random sampling and proportional allocation can be further examined for different sizes of the overall sample. For simple random sampling, one can express the variance of $\bar{y}$ as $V_S = S^2/n - S^2/43 = S^2/n - 31.68/43 = S^2/n - 0.74$. For the variance in (5.23) for proportional allocation, $\Sigma_1^G W_gS_g^2 = 9.68$, and $V_P = 9.68/n - 9.68/43 = 9.68 - 0.23$. Since $\Sigma W_gS_g = 2.19$ and $\Sigma_1^G W_gS_g^2 = 9.68$, for Neyman allocation, from (5.26), $V_N = 4.80/n - 0.23$.

Table 5.3 presents these variances and the relative precisions for Neyman allocation for $n = 5, 10, 15,$ and 20. The relative precisions are also presented in Figure 5.1. Notice from (5.24) that the precision of proportional allocation relative to simple random sampling is $V_S/V_P = 31.68/9.68 = 3.27$, and it is the same for any sample size.

Table 5.3. Variances for simple random sampling and proportional and Neyman allocations; relative precision of Neyman allocation.

Sample Size, n	V_S	V_P	V_N	V_S/V_N	V_P/V_N
4	7.180	2.190	0.965	7.4	2.3
6	4.540	1.383	0.567	8.0	2.4
8	3.220	0.980	0.368	8.8	2.7
10	2.428	0.738	0.248	9.8	3.0
12	1.900	0.577	0.168	11.3	3.4
14	1.523	0.461	0.111	13.7	4.1

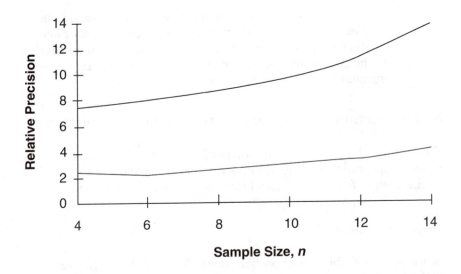

Figure 5.1. Precision of Neyman allocation relative to simple random sampling (top) and proportional allocation (bottom).

5.11   Summary on the precision of the allocations

The results from the comparisons of Sections 5.8 through 5.10 can be summarized as follows;

1. For a given sample size n, proportional and Neyman allocations estimate $\bar{Y}$ and Y with smaller variances than a simple random sample from the entire population.
2. Neyman allocation has much smaller variance than proportional allocation if S_g vary, and much smaller variance than simple random sampling if both $\bar{Y}_g$ and S_g vary. Information on the relative values of S_g is needed for Neyman allocation.
3. As $\bar{Y}_g$ vary, the precision of proportional allocation relative to simple random sampling increases.
4. An increase in n decreases the variances of the estimators for simple random sampling as well as for proportional and Neyman allocations and also for equal division of the sample size for the strata. As n increases, the precision of Neyman allocation relative to proportional allocation and simple random sampling increases, but it has no effect on the precision of proportional allocation relative to simple random sampling.

5. Distributing the sample size equally to the strata can be convenient in some situations. This procedure can have smaller variance than proportional allocation if it results in larger sample sizes for strata with larger sizes and standard deviations, as for Neyman allocation.

5.12   Sample size allocation to estimate proportions

To estimate the population proportion P or the total C, allocation of a given sample can be determined directly from the formulas in Section 5.7 or the results for the mean and total in Sections 5.8 and 5.9.

Proportional allocation

Since $n_g = nW_g$ for this allocation, from (5.20) or (5.23), the variance of $\hat{P}_{\text{st}}$ becomes

$$V_P(\hat{P}_{\text{st}}) = \frac{(1-f)}{n}\sum_1^G W_g \frac{N_g}{N_g - 1} P_g Q_g,\qquad (5.30)$$

which is approximately equal to $(1 - f)\Sigma_1^G W_g P_g Q_g / n$.

Following the analysis in Section 5.8, proportional allocation of the sample will have smaller variance for $\hat{P}_{\text{st}}$ than equal distribution if n_g is larger for the strata with larger $S_g^2 = N_g P_g Q_g/(N_g - 1)$, that is, for P_g closer to 0.5 than zero or unity.

Neyman allocation

For this procedure, as in the case of the mean, the variance in (5.20) is minimized for a given $n = \Sigma n_g$. If N_g does not differ much from $(N_g - 1)$, this minimization results in

$$\frac{n_g}{n} = \frac{W_g\sqrt{P_g Q_g}}{\sum W_g\sqrt{P_g Q_g}}.\qquad (5.31)$$

From (5.20) or (5.26), the minimum variance is approximately given by

$$V_N(\hat{P}_{\text{st}}) = \frac{\left(\sum W_g\sqrt{P_g Q_g}\right)^2}{n} - \frac{\sum W_g P_g Q_g}{N}.\qquad (5.32)$$

For exact expressions when N_g is not large, replace $P_g Q_g$ by $N_g P_g Q_g/(N_g - 1)$ in these equations.

Unlike proportional allocation, this method requires prior information regarding the unknown P_g. The sample sizes obtained from (5.31) do not differ much from proportional allocation even when P_g vary. For example, in the case of a population consisting of two strata, if $P_2 = 1 - P_1$, $n_1 = nW_1$ and $n_2 = nW_2$ from (5.25). Thus, the sample allocation in this case is identically the same as for proportional allocation, and the sample sizes do not depend on P_g. If valid prior information on the P_g is not available, proportional allocation can be recommended for practical situations.

5.13   Sample sizes to estimate means and totals

Sections 5.8 through 5.12, examined the distribution of the available overall sample size n for the strata. For practical situations, the sample size required for a specified criterion can be determined for each type of allocation.

To estimate $\bar{Y}_g$ and Y_g of the individual strata, one can find n_g needed with different criteria from the procedures in Sections 2.12. For proportional and Neyman allocations, procedures to determine the overall sample size n needed to estimate $\bar{Y}$ and Y with each of the following five criteria can be found:

1. The variance of $\hat{\bar{Y}}_{st}$ should not exceed a given value V.
2. The error of estimation $|\hat{\bar{Y}}_{st} - \bar{Y}|$ should not exceed a small positive quantity e except for a probability of α.
3. The relative error $|\hat{\bar{Y}}_{st} - \bar{Y}|/\bar{Y}$ should not exceed **a** except with probability α.
4. The confidence width for $\bar{Y}$ for a confidence probability of $(1 - \alpha)$ should not exceed a given value w.
5. The coefficient of variation of $\hat{\bar{Y}}_{st}$ should be smaller than C.

For proportional allocation, the solutions to these requirements are obtained by following the approaches in Section 2.12 with the expression in (5.23) for the variance. Now, for large N, the minimum sample sizes n_l required are obtained by multiplying $\Sigma W_g S_g^2$ by $(1/V)$, $(Z/e)^2$, $(Z/a\,\bar{Y})^2$, $(2Z/w)^2$, and $(1/C^2\,\bar{Y}^2)$ for the five criteria, respectively. Note that Z is the $(1 - \alpha)$ percentage point of the standard normal distribution, and if N is not very large, the sample size is given by $n = n_l/(1 + n_l/N)$.

Similarly, for Neyman allocation, the sample sizes are determined from the variance in (5.26). In this case, n_l is obtained by multiplying $(\Sigma W_g S_g)^2$ by the above factors and n is given by from $n_l/[1 + (n_l/N) \Sigma W_g S_g^2 / (\Sigma W_g S_g)^2]$.

Example 5.7. College tuitions: Consider estimating the average tuition for higher education for the colleges and universities in five states: New York, Pennsylvania, Massachusetts, New Jersey, and Connecticut. These states have relatively more institutions of higher education in the Northeast U.S. Baron's *Profiles of American Colleges* (1988) gives data on tuition, room and board expenses, and SAT (Scholastic Aptitude Test) scores for the educational institutions in the entire country.

The above five states together have $N = 329$ colleges and universities and they are classified as *Highly Competitive, Very Competitive*, and *Competitive*. These three categories consist of $N_1 = 55$, $N_2 = 74$, and $N_3 = 200$ institutions. Thus, the relative number of colleges in these three strata are $W_1 = 55/329 = 0.17$, $W_2 = 74/329 = 0.22$, and $W_3 = 200/329 = 0.61$.

From the above data, the averages $\overline{Y}_g$ of the tuitions in the three strata are found to be \$11.58, \$7.57, and \$6.13 thousand, respectively. The standard deviations S_g of the tuitions were \$0.77, \$1.24, and \$1.66 thousand, respectively. From the means of the strata, $\overline{Y} = (0.17)(11.58) + (0.22)(7.57) + (0.61)(6.13) = 7.37$. By using this as a preliminary estimate, sample size needed to estimate the mean for a specified S.E. can be found. For instance, consider the requirement that the S.E. of $\overline{Y}_{st}$ should not exceed 2.5% of the actual mean, that is, $7.37(0.025) = 0.18$ approximately.

From the above figures, $W_1 S_1^2 = 0.10$, $W_2 S_2^2 = 0.34$, and, $W_3 S_3^2 = 1.68$, and hence $\Sigma W_g S_g^2 = 2.12$. Similarly, $W_1 S_1 = 1.13$, $W_2 S_2 = 0.27$, $W_3 S_3 = 1.01$, and hence $\Sigma W_g S_g = 1.41$.

Now, for proportional allocation, $n_l = 2.12/(0.18)^2 = 65$ and $n = 65/(1 + 65/329) = 55$. The sample sizes for the strata are $n_1 = (0.17)(55) = 9$, $n_2 = (0.22)(55) = 12$, and $n_3 = (0.61)(55) = 34$.

Similarly, for Neyman allocation, $n_l = (1.41)^2/(0.18)^2 = 62$, and $n = 62/[1 + (62/329)(2.12)/(1.41)^2] = 52$. The sample sizes for the strata are $n_1 = (0.13/1.41)52 = 5$, $n_2 = (0.27/1.41)52 = 10$, and $n_3 = (1.01/1.41)52 = 37$.

5.14   Sample sizes to estimate proportions

As in the case of the mean and total, sample sizes to estimate P to satisfy the following requirements can be found:

1.   The variance of $\hat{P}_{st}$ should not exceed V.
2.   The margin of error $|\hat{P}_{st} - P|$ should be smaller than **e** except for a probability of α.

3. The relative error $|\hat{P}_{st} - P|/P$ should not exceed **a** except for a probability of α.
4. Confidence width for P with the confidence probability $(1 - \alpha)$ should not exceed w.
5. The coefficient of variation of $\hat{P}_{st}$ should be smaller than C.

For proportional allocation, from (5.30), the sample size n_l needed for the five prescriptions, respectively, are obtained by multiplying $\Sigma W_g P_g Q_g$ by $(1/V)$, $(Z/e)^2$, $(Z/aP)^2$, $(2Z/w)^2$ and $(1/CP^2)$, and $n = n_l/(1 + n_l/N)$.

For Neyman allocation, from (5.31), n_l is obtained by multiplying $(\Sigma W_g \sqrt{P_g Q_g})^2$ by the above factors, and $n = n_l/[1 + (n_l/N)(\Sigma W_g P_g Q_g)/(\Sigma W_g \sqrt{P_g Q_g})^2]$.

If the strata sizes N_g are not large, multiply $P_g Q_g$ by $N_g/(N_g - 1)$ in the above solutions. If valid prior information on P_g is not available, all the solutions can be obtained with $P_g = 0.5$.

Note that the same solutions are obtained for the estimation of the total count $C = NP$ with the corresponding prescriptions. For example, the first solution is obtained for estimating C with a variance not exceeding $N^2 V$.

Example 5.8. Prescribed S.E: For the energy consumption in Example 5.3, determine the sample size to estimate the proportion with the S.E. not exceeding 10%. Since $W_1 = 10/25$, $W_2 = 15/25$, $P_1 = 7/10$, and $P_2 = 8/15$.

From these figures, $W_1 P_1 Q_1 = 0.084$, $W_2 P_2 Q_2 = 0.1493$, and hence $\Sigma W_g P_g Q_g = 0.2333$. For proportional allocation, $n_l = 0.2333/0.10 = 24$ approximately, and $n = 24/[1 + (24/25)] = 12.2$ or 13 approximately. The sample sizes for the strata are $n_1 = 13(0.4) = 5$ and $n_2 = 13(0.6) = 8$.

Further, $W_1(P_1 Q_1)^{1/2} = 0.1833$, $W_2(P_2 Q_2)^{1/2} = 0.2993$, and $\Sigma W_g \sqrt{P_g Q_g} = 0.4826$. Now, for Neyman allocation, $n_l = (0.4826)^2/(0.10) = 23.29$ or 24 approximately. Now, $n = 1/[1 = (24/25)(0.2333/0.2329) = 12.23$ or 13 approximately. The sample sizes for the strata are $n_1 = 13(0.1833/0.4826) = 5$ and $13(0.2993/0.4826) = 8$.

5.15   Sample sizes for minimizing variance or cost

In the last two sections, expenses for collecting information from the sample units were not considered. The costs for obtaining information from a unit can be different for the strata. A cost function as suggested in the literature takes the form

$$E = e_0 + e_1 n_1 + \cdots + e_G n_G = e_0 + \Sigma e_g n_g, \qquad (5.33)$$

see Cochran (1977, Chap. 5), for example. In this linear function, e_0 represents the initial expenses for making arrangements for the survey, which may be ignored for some established surveys, e_g is the cost for interviewing a selected unit in the gth stratum, and E is the total allowable expenditure.

As outlined in Appendix A5, the sample sizes needed to minimize the variance in (5.13) for a given cost E, or for minimizing the cost in (5.33) for a given value V of the variance are obtained from

$$\frac{n_g}{n} = \frac{N_g S_g / \sqrt{e_g}}{\sum (N_g S_g / \sqrt{e_g})} = \frac{W_g S_g / \sqrt{e_g}}{\sum (W_g S_g \sqrt{e_g})}. \tag{5.34}$$

For the variance or the cost to be minimum, this optimum allocation suggests that larger sample sizes are needed for strata with larger sizes and standard deviations, and they can be large if the costs of sampling are not large. If e_g are the same for all the strata, (5.34) coincides with the Neyman allocation in (5.25).

Substituting (5.34) in (5.13), the minimum of the variance is

$$V_{\text{opt}}(\bar{y}_{st}) = \frac{\sum W_g S_g \sqrt{e_g} \sum (W_g S_g / \sqrt{e_g})}{n} - \frac{\sum W_g S_g^2}{N}. \tag{5.35}$$

For a given V, n is obtained from this expression. Similarly, substituting (5.34) in (5.33), the minimum cost is

$$E = e_0 + \frac{\sum W_g S_g \sqrt{e_g}}{\sum (W_g S_g / \sqrt{e_g})} n. \tag{5.36}$$

The sample size n for the available budget is obtained from this expression.

Example 5.9. Optimum allocation: For the illustration in Example 5.7, consider an initial expense of $e_0 = \$400$, and the expenses of $e_1 = \$50$, $e_2 = \$50$, and $e_3 = \$45$ to obtain information on a unit on tuition from the institutions in the three strata. With these costs, $W_1 S_1 \sqrt{e_1} = 0.92$, $W_2 S_2 \sqrt{e_2} = 1.93$, and $W_3 S_3 \sqrt{e_3} = 6.80$, and hence $\sum W_g S_g \sqrt{e_g} = 9.65$. Similarly, $W_1 S_1 / \sqrt{e_1} = 0.0184$, $W_2 S_2 / \sqrt{e_2} = 0.0386$, and $W_3 S_3 / \sqrt{e_3} = 0.1510$, and hence $\sum (W_g S_g / \sqrt{e_g}) = 0.2079$.

If the S.E. of $\hat{Y}_{st}$ should not exceed 0.18 as in Example 5.7, from (5.35), $9.65(0.21)/n - 2.13/329 = (0.18)^2$, and hence $n = 53$. The allocation in (5.34) results in $n_1 = (0.0184/0.2079)53 = 5$, $n_2 = (0.0386/0.2079)53 = 10$, and $n_3 = (0.1510/0.2079)53 = 38$. From (5.36), the total cost for the survey is equal to $2860.

On the other hand, if the total cost should not exceed $3000, for example, from (5.36), $400 + (9.65/0.2079)n = 3000$, and hence $n = 56$. Now, from (534), $n_1 = (0.0184/0.2079)56 = 5$, $n_2 = (0.0386/0.2079)56 = 10$, and $n_3 = (0.1510/0.2079)56 = 41$. In this case, the variance for $\hat{Y}_{st}$ from (5.35) is equal to 0.0293, and hence S.E.$(\hat{Y}_{st}) = 0.17$. With the increase of $3000 - 2860 = 140$, S.E.$(\hat{Y}_{st})$ is reduced by 0.01, that is, by $0.01/0.18$, or about 6%.

5.16   Further topics

Poststratification

The observations of a simple random sample of the students of a college can be classified into the freshman, sophomore, junior, and senior classes. The returns of a mail survey on the physicians in a region can be categorized according to their specialties. The responses of the public to a marketing or political survey can be classified into strata defined according to one or more of the characteristics such as age, sex, profession, family size, and income level.

Means, totals, and proportions of the above type of strata can be estimated from the sample units *observed* in the strata in a sample of size n drawn from the N population units. Since the sizes n_g of the observed samples are not fixed and they are random, there will be some loss in precision for the estimators if average variances are considered. This procedure, known as poststratification, is described below for estimating the means of the strata and the population.

The mean $\bar{y}_g$ of the n_g observed sample units can be used to estimate the stratum mean $\bar{Y}_g$. For an observed n_g, the mean $\bar{y}_g$ and variance s_g^2 are unbiased for $\bar{Y}_g$ and S_g^2. Further, $V(\bar{y}_g \mid n_g) = (N_g - n_g)S_g^2/N_g n_g$, which can be estimated from replacing S_g^2 by s_g^2. By replacing n_g by its expectation nW_g, an approximation to the average of this variance becomes $(1 - f)S_g^2/nW_g$.

With poststratification, the estimator of the population mean is

$$\hat{Y}_{pst} = \sum_1^G W_g \bar{y}_g, \tag{5.37}$$

which is unbiased for $\bar{Y}$. Since $V(\hat{\bar{Y}}_{pst}) = \Sigma W_g^2 V(\bar{y}_g)$, replacing n_g by $n W_g$ as above, the average variance of $\hat{\bar{Y}}_{pst}$ is approximately given by $(1 - f) \Sigma W_g S_g^2 / n$, which is the same as the variance in (5.23) for proportional allocation. Thus, for large sample sizes, $\hat{\bar{Y}}_{pst}$ has the same precision as $\hat{\bar{Y}}_{st}$ with proportional allocation of the sample.

Estimating differences of the stratum means

With samples of sizes n_1 and n_2 selected from two strata, an unbiased estimator of the difference of their means $(\bar{Y}_1 - \bar{Y}_2)$ is given by $(\bar{y}_1 - \bar{y}_2)$, which has the variance of $(N_1 - n_1) S_1^2 / n_1 + (N_2 - n_2) S_2^2 / n_2$. As shown in Appendix A5, for given $n = n_1 + n_2$, this variance is minimized if n_1 and n_2 are proportional to S_1 and S_2, respectively, that is, $(n_1/n) = S_1/(S_1 + S_2)$ and $(n_2/n) = S_2/(S_1 + S_2)$. In general, this division of the sample is not the same as the proportional or Neyman allocations described in Sections 5.8 and 5.9. However, it becomes the same as Neyman allocation if $N_1 = N_2$ and the same as proportional allocation if $S_1 = S_2$ in addition.

Two-phase sampling for stratification

When the sizes N_g are not known, Neyman (1938) suggests this procedure, which is also known as **double sampling** for stratification. In this procedure, an initial sample of large size n' is selected from the entire population and n_g', $g = (1, 2,...,G)$, $\Sigma n_g' = n'$, units of the gth stratum are identified. Notice that $w_g = n_g'/n'$ is an unbiased estimator for $W_g = N_g/N$. At the second phase, samples of size n_g are selected from the n_g' observed units, the means and variances $(\bar{y}_g, s_g^2)$ are obtained, and the population mean $\bar{Y}$ is estimated from $\Sigma w_g \bar{y}_g$.

   For repeated sampling at both phases, Cochran (1977, pp. 327–335) and J.N.K. Rao (1973) present expressions for the variance of $\Sigma w_g \bar{y}_g$ and the estimator for the variance. These authors and Treder and Sedransk (1993) describe procedures for determining the sample sizes n' and n_g.

Sample size determination for two or more characteristics

Several surveys are usually conducted to estimate the population quantities such as the means, totals, and proportions of more than one characteristic. Proportional or equal distribution of the available sample size can be considered for such surveys.

Neyman allocation based on the different characteristics can result in different allocations of the overall sample size. Yates (1960) suggests finding the sample sizes by minimizing the weighted average of the variances in (5.13) for all the characteristics; the weights are determined from the importance of the characteristics. Cochran (1977, pp. 119–123) averages the Neyman allocations of the different characteristics. Chatterjee (1968) and Bethel (1989) suggest alternative procedures for allocating the sample size.

Strata formation

Sections 1.8 and 5.1 presented illustrations of different types of strata considered for different purposes. If the main purpose of stratification is to estimate the population quantities such as the mean with high precision, as noted in Sections 5.8 and 5.9, the units within each stratum should be close to each other but the means of the strata should differ as much as possible. Ekman (1959) and Dalenius and Hodges (1959) suggest procedures for formation of the strata to estimate the mean or total. Cochran (1961) examines the difference procedures available for the division of a population into strata.

For the illustration in Example 5.1 to estimate rice production, the strata based on the levels of rice production have been constructed. To estimate the average tuition in Example 5.7, the different levels of competitiveness were considered to form the strata. In this case, an initial attempt to consider five U.S. states as the strata resulted in almost the same means and variances for all the strata. The reason for this outcome was that each state has some educational institutions with high and some with low tuitions.

Exercises

5.1. With the sample information in Example 5.1, (a) estimate the differences of the means and totals of the rice productions in the two strata, (b) find the S.E. of the estimates, and (c) find the 95% confidence limits for the differences.

5.2. In samples of size three and five selected from the two strata of Example 5.3, respectively, units (1, 6, 24) and (4, 10, 12, 15, 20) appeared. For the average and total consumption of the 25 countries, find (a) the estimates, (b) their S.E., and (c) the 95% confidence limits.

5.3. For the estimators of the average rice production in Example 5.1 that can be obtained through proportional and Neyman allocations of a sample of size $n = 16$, (a) find the variances and (b) compare their precisions relative to the mean of a simple random sample of the same size.

5.4. A total of 209 students responded to the survey conducted by the students in 1999, described in Section 1.8. The number of freshman, sophomore, junior, and senior students responded to the survey were 50, 30, 56, and 73. Among these students, 82, 80, 57, and 60%, respectively, had part-time employment on campus, about 11 hours a week for each group. The university consisted of 807, 798, 966, and 1106 students in the four classes, a total of 3677. For the 1605 freshman-sophomore and the 2072 junior–senior classes, find (a) estimates for the percentages having part-time employment and (b) The standard errors of the estimates. (c) Estimate the difference of the percentages for the two groups and find its S.E.

5.5. Divide the population of the 32 units in Table T2 in the Appendix into two strata with the math scores below 600 and 600 or more. For proportional and Neyman allocations of a sample of size 10, compare the variances of the stratified estimators for the means of the total (verbal + math) scores.

5.6. With the stratification in Exercise 5.5 and a sample of size 10, compare the variances of proportional and Neyman allocations for estimating the proportion of the 32 candidates scoring a total of 1100 or more.

5.7. For the 200 managers and 800 engineers of a corporation, the standard deviations of the number of days a year spent on research were presumed to be 30 and 60 days, respectively. Find the sample size needed for proportional allocation to estimate the population mean with the S.E. of the estimator not exceeding 10 and its allocation for the two groups.

5.8. With the information in Exercise 5.7, (a) find the sample size required if the minimum variance of the estimator for the difference of the means of the two groups should not exceed 200, and (b) find the distribution of the sample size for the groups.

5.9. Using the information in Example 5.7, find the sample sizes needed for proportional and Neyman allocations to estimate the average tuition if the width of the 95% confidence limits should not exceed $1000.

5.10. To estimate the population proportion in Example 5.4 with an error not exceeding 12% except with a probability of 0.05, find the sample sizes required for proportional and Neyman allocations.

5.11. To estimate the average tuition in Example 5.9 when $e_0 = 500$, $e_1 = 50$, $e_2 = 100$, and $e_3 = 150$ and $5000 is available for the survey, find the sample sizes for the strata and the minimum variance and S.E. of the estimator that can be obtained with these sample sizes.

5.12. As mentioned in Example 5.7, the standard deviations of the tuitions for the three types of colleges were $0.77, $1.24, and $1.66 thousand. Similarly, for the expenses for room and board (R&B), the standard deviations were $0.61, $0.67, and $0.83 thousand. (a) Find the standard errors of the estimators for the averages of the tuition and R&B expenses for a sample of size 50 allocated proportionally to the three types of colleges. (b) For the tuition, distribute the 50 units among the three types through Neyman allocation and find the S.E. of the estimator. (c) Similarly, distribute the 50 units among the three types for R&B through Neyman allocation and find the S.E. of the estimator.

5.13. *Project.* Consider the first five and next five states of Table T4 in the Appendix as two strata. From all the samples of size three selected independently from each of the strata, estimate (a) average of private enrollments, (b) difference of the averages of public and private enrollments, and (c) difference between the averages of the two strata for private enrollments. From the above estimates, find the expectations and variances of the corresponding estimators, and verify that they coincide with the exact expressions for the expectations and variances. Note that the expectations and variances for (a) and (b) can be obtained from the $_5C_3 + {}_5C_3 = 20$ estimates, but they should be obtained from the $_5C_3 \times {}_5C_3 = 100$ estimates.

5.14. Divide 47 of the states in Table T3 in the Appendix into two strata consisting of the 25 states with 2000 or more persons over 25 years old and the remaining 22 states.

(a) To estimate the average number of persons over 25 in the 47 states, find the S.E. of the stratified estimator for samples of size 5 from each of the strata. (b) For Neyman allocation of the sample of size 10, find the S.E. of the above estimator. (c) To estimate the above average for all the 51 entries, describe the procedure for combining the figures for the four largest states with the sample estimates in (a) or (b), and find the standard errors of the resulting estimators.

5.15.   Express the sample variance s^2 obtained from a sample of n units as in (5.6) for population variance S^2.

5.16.   For the observed sample size n_g in the gth stratum from a sample of n units from the entire population, show that $E(1/n_g) \geqslant 1/E(n_g) = nW_g$ (a) from the Cauchy–Schwartz inequality in Appendix A3 and (b) noting that the covariance of n_g and $1/n_g$ is negative. (c) Use this result to show that the approximation to the average variance of the poststratified estimator presented in Section 5.16 is smaller than the actual average.

Appendix A5

An expression for the variance

The numerator of (5.5) can be expressed as

$$\sum_{1}^{G Ng}\sum_{1}(y_{gi} - \bar{Y})^2 = \sum_{1}^{G}\sum_{1}^{Ng}[(y_{gi} - \bar{Y}_g) + (\bar{Y}_g - \bar{Y})]^2$$

$$= \sum_{1}^{G}\sum_{1}^{Ng}(y_{gi} - \bar{Y}_g)^2 + \sum_{1}^{G}N_g(\bar{Y}_g - \bar{Y})^2$$

$$= \sum_{1}^{G}(N_g - 1)S_g^2 + \sum N_g(\bar{Y}_g - \bar{Y})^2.$$

The cross-product terms of the first expression on the right-hand side vanish since $\sum_{1}^{Ng}(y_{gi} - \bar{Y}_g) = 0$. The first and second terms of the last two expressions are the *within SS* and *between SS* (sum of squares). These two terms are the expressions for the variations within the strata and among the means of the strata, respectively.

Variance of the estimator for the mean

From (5.11), the variance of $\hat{\bar{Y}}_{st}$ is

$$V(\hat{\bar{Y}}_{st}) = \sum_g W_g^2 V(\bar{y}_g) + \sum_g \sum_{g \neq j} W_g W_j \, \mathrm{Cov}(\bar{y}_g, \bar{y}_j).$$

Since samples are selected independently from the strata, $E(\bar{y}_g \bar{y}_j) = E(\bar{y}_g)E(\bar{y}_j)$ and hence the covariance on the right side vanishes.

Neyman allocation

To find n_g that minimizes the variance in (5.13) for a given n, let

$$\Delta = \sum W_g^2 \left(\frac{1}{n_g} - \frac{1}{N_g} \right) S_g^2 + \lambda \left(\sum n_g - n \right),$$

where λ is the Lagrangian multiplier. Setting the derivative of Δ with respect to n_g to zero, $n_g = W_g S_g / \sqrt{\lambda}$. Since $n = \sum n_g = \sum W_g S_g / \sqrt{\lambda}$, the optimum value of n_g is given by (5.25).
   This result can also be obtained from the Cauchy–Schwartz inequality in Appendix A3 by writing $a_g = W_g S_g / (n_g)^{1/2}$ and $b_g = (n_g)^{1/2}$.

Optimum allocation

The sample size required for minimizing the variance in (5.13) for a given cost is obtained by setting the first derivative of

$$\Delta_1 = V(\hat{\bar{Y}}_{st}) + \lambda \left(E - e_0 - \sum e_g n_g \right)$$

with respect to n_g to zero. Similarly, the sample size to minimize the cost for a specified value V of the variance can be found by setting the first derivatives of

$$\Delta_2 = e_0 + \sum e_g n_g + \lambda [V(\hat{\bar{Y}}_{st}) - V]$$

to zero. The solution for both these procedures is given by (5.34).
   The solution can also be obtained from the Cauchy-Schwartz inequality by writing $a_g = W_g S_g / (n_g)^{1/2}$ and $b_g = (e_g n_g)^{1/2}$.

CHAPTER 6

Subpopulations

6.1   Introduction

From the observations of a sample drawn from a population or region, estimates for the totals, means, and proportions for specified groups, **subpopulations, domains**, and areas can also be obtained. Households classified according to their sizes, income levels, and age or educational level of the head of the household are examples of subpopulations. Industries classified according to the employee sizes, production levels, and profits provide other examples of these groups. Certain number of units of each of the subpopulations or groups will appear in a sufficiently large sample selected from the entire population.

Political polls provide several illustrations of estimating percentages for subpopulations. For instance, in the local and national elections for public offices, including those of the president, governors, senators, representatives, and mayors, polls conducted before and on election day are classified into several categories. The classifications are usually based on age, education, income, sex, marital status, political affiliation, race, religion, and other demographic and socioeconomic characteristics of the voters. Percentages of the voters favoring each of the candidates are estimated for the entire nation as well as for each of the classifications.

In the studies related to specified subpopulations, importance is given to the estimation for each subpopulation and differences between the subpopulations. If the population is initially stratified, estimates for a subpopulation of interest are obtained from its units appearing in the samples of each of the strata. These estimates can also be obtained from the poststratification described in Section 5.16.

Estimation procedures for the totals and means of subpopulations from a simple random or stratified sample are presented in the following sections. For **small areas**, subpopulations, and regions of small sizes, procedures for increasing the precision of the estimators are described in Chapter 12.

6.2　Totals, means, and variances

Consider a population with k groups with N_i observations, $N = \Sigma_1^k N_i$, in the ith group. Let y_{ij}, $j = 1,...,N_i$, represent the jth observation in the ith group. The total, mean, and variance of the ith group are

$$Y_i = \sum_1^{N_i} y_{ij} = y_{i1} + y_{i2} + \cdots + y_{iN_i}, \qquad (6.1)$$

$$\bar{Y}_i = \frac{Y_i}{N_i} \qquad (6.2)$$

and

$$S_i^2 = \frac{\sum_1^{N_i}(y_{ij} - \bar{Y}_i)^2}{N_i - 1}. \qquad (6.3)$$

6.3　Estimation of the means and their differences

When a sample of size n is selected without replacement from the N population units, the probability of observing $(n_1, n_2, ..., n_k)$ units, $n = \Sigma_1^k n_i$, in the groups is

$$P(n_1, n_2, ..., n_k) = \frac{\binom{N_1}{n_1}\binom{N_2}{n_2}\cdots\binom{N_k}{n_k}}{\binom{N}{n}}. \qquad (6.4)$$

Note that conditional on n_i, the probability for a unit from the ith group to appear in the sample is n_i/N_i and the probability for a specified pair of units from the ith group to be selected into the sample is $n_i(n_i - 1)/N_i(N_i - 1)$.

Means

The sample mean and variance for the ith group are

$$\bar{y}_i = \frac{1}{n_i}\sum^{n_i} y_{ij}. \qquad (6.5)$$

and

$$s_i^2 = \frac{\sum_1^{n_i}(y_{ij} - \bar{Y}_i)^2}{n_i - 1}. \tag{6.6}$$

For an observed n_i, $\bar{y}_i$ is unbiased for $\bar{Y}_i$ and has the variance

$$V(\bar{y}_i) = \frac{N_i - n_i}{N_i n_i} S_i^2. \tag{6.7}$$

From the probability distribution in (6.4), $E(n_i) = nW_i$, where $W_i = N_i/N$ is the proportion of the units in the ith domain. Substitution of this expectation for n_i provides an approximation to the average of the above variance.

An estimator of the above variance is given by

$$v(\bar{y}_i) = \frac{N_i - n_i}{N_i n_i} s_i^2. \tag{6.8}$$

If N_i is not known, it may be replaced by $\hat{N}_i = Nw_i = N(n_i/n)$, where w_i is the sample proportion of the units observed in the ith group. In this case (6.8) becomes $(1 - f)s_i^2/n_i$.

Example 6.1. Hospital expenditures: In the American Hospital Association *Guide to the Health Care Field* (1988), complete information on the following characteristics is available for 217 short-term hospitals (average length of stay of the patients is less than 30 days) in New York State with general medical and surgical services: (1) Number of beds, (2) average number of patients treated per day, (3) rate of occupancy of the beds, and (4) total annual expenses. Among the 217 hospitals, 150 were owned by individuals, partnerships, or churches. The remaining 67 were operated by federal, state, local or city governments.

A sample of 40 from the 217 hospitals has been selected randomly without replacement. In this sample, 29 hospitals of the first type and 11 of the second type appeared. The means and standard deviations of all the 40 hospitals and the samples observed in the two groups are presented in Table 6.1.

As can be seen, the estimate of the average amount of expenses for the first group is $\bar{y}_1 = \$61.1$ million. With the known value of $N_1 = 150$ and the sample size $n_1 = 29$, from (6.8), $v(\bar{y}_1) = 106.6$ and S.E.$(\bar{y}_1) = 10.3$.

Table 6.1. Means (top) and standard deviations (bottom) for a sample of 40 hospitals.

	Beds	Patients	Occupancy Rates	Expenses, (millions $)
Entire				
sample	339.7	289.7	81.1	69.2
$n = 40$	227.2	201.5	11.6	65.0
Group 1	325.9	276.5	81.7	61.1
$n_1 = 29$	234.8	206.9	11.7	61.9
Group 2	376.1	324.5	79.6	90.7
$n_2 = 11$	211.9	191.5	11.9	71.2

Source: American Hospital Association, *Guide to the Health Care Field* (1988).

If N_1 is not known, its estimate is $\hat{N}_1 = 217(29/40) = 157$. Substituting this value for N_1 in (6.8), $v(\bar{y}_1) = 112.7$ and S.E.$(\bar{y}_1) = 10.6$.

Difference between two groups

An unbiased estimator of the difference $(\bar{Y}_i - \bar{Y}_j)$ of the means of two groups is $(\bar{y}_i - \bar{y}_j)$ and its variance is given by

$$V(\bar{y}_i - \bar{y}_j) = (1 - f_i)\frac{S_i^2}{n_i} + (1 - f_j)\frac{S_i^2}{n_j}, \qquad (6.9)$$

where $f_i = n_i/N_i$ and $f_j = n_j/N_j$. For an estimate of this variance, replace the group variances S_i^2 and S_j^2 by their sample estimates s_i^2 and s_j^2. If the group sizes N_i and N_j are not known, they are replaced by the estimates $n_i(N/n)$ and $n_j(N/n)$, that is, f_i and f_j are replaced by the common sampling fraction $f = n/N$.

From Table 6.1, we find that an estimate for the increase $(\bar{Y}_2 - \bar{Y}_1)$ in the average expenditure for the second group from that of the first is $(\bar{y}_2 - \bar{y}_1) = 90.7 - 61.1 = \$29.6$ million. From (6.9), an estimate of $V(\bar{y}_2 - \bar{y}_1)$ is given by $v(\bar{y}_2 - \bar{y}_1) = (67 - 11)(71.2)^2/(67 \times 11) + (150 - 29)(61.9)^2/(150 \times 29) = 491.78$, and hence S.E. $(\bar{y}_2 - \bar{y}_1) = 22.18$.

As seen in Example 6.1, $\hat{N}_1 = 157$ and hence $\hat{N}_2 = 60$. With these estimated sizes, $v(\bar{y}_2 - \bar{y}_1) = 484$ and hence S.E.$(\bar{y}_2 - \bar{y}_1) = 22$.

6.4 Totals of subpopulations

Known sizes

In several applications, it is often of interest to estimate the totals of subpopulations. If N_i is known, an unbiased estimator of Y_i is

$$\hat{Y}_i = N_i \bar{y}_i = \frac{N_i}{n_i} \sum_1^{n_i} y_{ij}. \tag{6.10}$$

For a given n_i, its variance is obtained from

$$V(\hat{Y}_i) = N_i^2 V(\bar{y}_i) = \frac{N_i(N_i - n_i)}{n_i} S_i^2. \tag{6.11}$$

For estimating this variance, substitute s_i^2 for S_i^2.

Estimated sizes

If N_i is not known, it can be estimated from $\hat{N}_i = N(n_i/n)$. With this estimator, from (6.10), an estimator for the total is

$$\hat{\hat{Y}}_i = \hat{N}_i \bar{y}_i = \frac{N}{n} \sum_1^{n_i} y_{ij}. \tag{6.12}$$

Since

$$E(\hat{\hat{Y}}_i \mid n_i) = \hat{N}_i \bar{Y}_i, \tag{6.13}$$

$$E(\hat{\hat{Y}}_i) = E(\hat{N}_i) \bar{Y}_i = N_i \bar{Y}_i = Y_i. \tag{6.14}$$

For a given n_i, the variance of this unbiased estimator is

$$V(\hat{\hat{Y}}_i) = \hat{N}_i^2 V(\bar{y}_i) = \frac{N^2 n_i (N_i - n_i)}{N_i n^2} S_i^2. \tag{6.15}$$

If n_i is replaced by its expected value nW_i, approximations to the averages of both (6.11) and (6.15) will be equal to $N(N - n)W_i S_i^2/n$.

Notice that when N_i is known, the estimator in (6.10) is obtained by dividing the sample total by n_i/N_i, which is the probability of a unit of the ith group appearing in a sample of size n_i from the N_i units of the ith group. If N_i is not known, as in (6.12), the estimator is obtained by dividing the sample total by n/N, which is the probability of selecting a unit in the sample of size n from the entire population.

Example 6.2. Hospital expenditures: From the data in Table 6.1, with the known value of $N_1 = 150$, the estimate for the total expenses of the first group is $\bar{Y}_1 = 150(61.1) = \$9165$ million, that is, close to \$9.2 billion. From Example 6.1, S.E. $(\bar{y}_1) = 10.3$, and hence S.E.$(\hat{Y}_1) = 150(10.3) = \$1545$ million.

If N_1 is not known, with $\hat{N}_1 = 157$, $(\hat{\hat{Y}}_1) = 157(61.1) = \$9592.7$ milion, that is, about \$9.6 billion. Since S.E.$(\bar{y}_1) = 10.6$ for this case, S.E.$(\hat{Y}_1) = 157(10.6) = \$1664.2$ million.

6.5   Sample sizes for estimating the means and totals

The overall sample size n for a survey can be determined for estimating the means or totals of subpopulations with required precision. If it is required that the variance of $\bar{y}_i$ should not exceed V_i, $i = 1,...,k$, substituting $E(n_i) = nW_i$ for n_i in (6.7),

$$\left(\frac{1}{n} - \frac{1}{N}\right)\frac{S_i^2}{W_i} \le V_i. \tag{6.16}$$

Solving this equation with the equality sign, n is obtained from $n_m/(1 + n_m/N)$, where n_m is the maximum value of S_i^2/V_iW_i.

Example 6.3. Sample size for hospital expenditures: The proportions of the units in the two groups of the previous section are $W_1 = 150/217 = 0.69$ and $W_2 = 0.31$. If it is of interest to estimate the average expenses of these groups with standard errors not exceeding 15 and 30, using the sample estimates 61.9 and 71.2 for S_1 and S_2.

$$n_m = \text{Max}\left[\frac{(61.9)^2}{(15)^2(0.69)}, \frac{(71.2)^2}{(30)^2(0.31)}\right] = 25.$$

The required sample size n is equal to $25/(1 + 25/217) = 23$.

Similar methods can be used to find the overall sample size with prescriptions on the errors of estimation, coefficients of variations, and confidence widths of the group means. The same type of approaches can also be used to estimate the differences of group means. Sedransk (1965, 1967), Booth and Sedransk (1969), and Liao and Sedransk (1983) suggest procedures for determining the sample sizes with the above type of constraints.

6.6 Proportions and counts

Consider an attribute with two classes and k subpopulations. Among the N_i units of the ith subpopulation, $i = 1, 2, ..., k$, let C_i and $(N_i - C_i)$ of the units belong to the two classes. In a random sample of size n from the population, n_i units will be observed in the ith subpopulation, with c_i and $(n_i - c_i)$ units from the two classes. Table 6.2 describes the compositions of the population and sample.

The sample proportion $p_i = c_i/n_i$ is unbiased for $P_i = C_i/N_i$. Its variance and estimator of variance are given by

$$V(p_i) = \frac{N_i - n_i}{N_i - 1} \frac{P_i(1 - P_i)}{n_i} \tag{6.17}$$

and

$$v(p_i) = \frac{N_i - n_i}{N_i} \frac{p_i(1 - p_i)}{n_i - 1}. \tag{6.18}$$

An unbiased estimator of C_i is $\hat{C}_i = N_i p_i$. If the size of the subpopulation N_i is not known, an unbiased estimator of the total is given by $\hat{N}_i p_i$ where $\hat{N}_i = N(n_i/n)$.

For the case of more than two classes, estimates and their standard errors for the proportions and totals in the different subpopulations

Table 6.2. Household composition.

| | Residences with Number of Persons | | | | | | |
	1	2	3	4	5	6+	Total
Owners	58	89	46	29	15	5	242
Renters	42	58	33	18	4	3	158

and their differences can be found through suitable modifications of the expressions in Section 4.7.

Example 6.4. Household composition: The figures in Table 6.2 refer to the numbers of persons in a random sample of 400 from the 14,706 residences in a town in New York State. During 1 year, the town assessor reported that 8549 of the houses were occupied by owners and the remaining 6157 were rented.

An estimate of the proportion of the four-member households in the town is $(29 + 18)/400 = 0.1175$. From (6.18), the variance of this estimate is $(1,4706 - 400)(0.1175)(0.8825)/(1,4706 \times 399) = 2.53 \times 10^{-4}$ and hence its S.E. is equal to 0.016. An estimate for the total number of houses of this type is $14,706(0.1175) = 1728$ and its S.E. is $14,706(0.016) = 235$.

For the owners, the estimates of the above proportion is $29/242 = 0.1198$. With the known size of 8549 for this group, from (6.18), the variance of this estimate is $(8549 - 242)(0.1198)(0.8802)/(8549 \times 241) = 4.25 \times 10^{-4}$ and hence its S.E. is equal to 0.021. If the size is unknown, its estimate is $14,706(242/400) = 8897$. Substituting this figure in (6.18), the S.E. again is close to 0.021.

An estimate for the total number of households of the above type among the owner-occupied residences is $8549(0.1198) = 1024$. The S.E. of this estimate is $8549(0.021) = 180$.

For the owner-occupied residences, an estimate of the difference in the proportions of the two-member and four-member households is $(89/242) - (29/242) = 0.248$. The variance of this estimate is $[(8549 - 242)/8549][0.2325 + 0.1054 + 2(0.0441)]/241 = 0.0017$, and hence its S.E. equals 0.041. An estimate of the difference in the numbers of houses of these types is $8549 (0.248) = 1624$, and the S.E. of this estimate is $8549 (0.041) = 351$.

For finding the sample sizes to estimate the proportions of qualitative characteristics in the subpopulations or differences between the proportions, the procedures in Section 6.5 can be used, by replacing S_i^2 with $N_i P_i (1 - P_i)/(N_i - 1)$ or approximately by $P_i(1 - P_i)$.

6.7   Subpopulations of a stratum

Estimates for specified subpopulations can be obtained from the observations of a stratified random sample. Estimates for the subpopulations from a single stratum are described in this section. Section 6.8 presents estimates for the subpopulations obtained from all the strata.

Let $N_{gj}, j = 1,...,k, \Sigma_j N_{gj} = N_g$, denote the size of the jth subpopulation in the gth stratum. The observations of these subpopulations can be denoted by $y_{gij}, i = 1,...,N_{gj}$. The mean and variance of this group are

$$\bar{Y}_{gj} = \frac{\sum_1^{N_{gj}} y_{gij}}{N_{gj}} \quad \text{and} \quad S_{gj}^2 = \frac{\sum_1^{N_{gj}} (y_{gij} - \bar{Y}_{gj})^2}{N_{gj} - 1}. \tag{6.19}$$

In a random sample of size n_g from the N_g units of the stratum, n_{gj} units, $\Sigma_j N_{gj} = n_g$, of the jth group will be observed. The sample mean and variance for this group are

$$\bar{y}_{gj} = \frac{\sum_1^{n_{gj}} y_{gij}}{N_{gj}} \quad \text{and} \quad s_{gj}^2 = \frac{\sum_1^{n_{gj}} (y_{gij} - \bar{y}_{gj})^2}{n_{gj} - 1}, \tag{6.20}$$

which are unbiased for $\bar{Y}_{gj}$ and S_{gj}^2. The variance of $\bar{y}_{gj}$ and its estimator are

$$V(\bar{y}_{gj}) = \frac{N_{gj} - n_{gj}}{N_{gj} n_{gj}} S_{gj}^2 \tag{6.21}$$

and

$$v(\bar{y}_{gj}) = \frac{N_{gj} - n_{gj}}{N_{gj} n_{gj}} s_{gj}^2. \tag{6.22}$$

An approximation to the average of (6.21) is obtained by replacing n_{gj} by its expectation $n_g(N_{gj}/N_g)$. If N_{gj} is not known, it is replaced by $\hat{N}_{gj} = (n_{gj}/n_g)N_g$ for estimating the variance.

Example 6.5. Athletic facility usage by the males and females of three strata: As an illustration, consider the 4000 undergraduate students, 1000 graduate students, and 200 faculty of a university as the three strata.

The first stratum consists of $N_{11} = 2000$ male students with the total Y_{11}, mean $\bar{Y}_{11}$, and variance S_{11}^2 for a characteristic of interest. This stratum has $N_{12} = 2000$ female students with total Y_{12}, mean $\bar{Y}_{12}$, and variance S_{12}^2. Similarly, for the $N_{21} = 600$ male and $N_{22} = 400$ female graduate students, these quantities are represented by $(Y_{21}, \bar{Y}_{21}, S_{21}^2)$ and $(Y_{22}, \bar{Y}_{22}, S_{22}^2)$, respectively. For the $N_{31} = 150$ and $N_{32} = 50$ female faculty, they can be represented by $(Y_{31}, \bar{Y}_{31}, S_{31}^2)$ and $(Y_{32}, \bar{Y}_{32}, S_{32}^2)$, respectively.

Table 6.3. Usage of athletic facilities.

	Males	Females	Strata and Sample Sizes
Undergraduates			
N_{1j}	2000	2000	4000
n_{1j}	5	5	10
$\bar{y}_{1j}$	3.4	1.8	
s_{1j}^2	1.3	0.7	
Graduates			
N_{2j}	600	400	1000
n_{2j}	3	2	5
$\bar{y}_{2j}$	4.67	2.5	
s_{2j}^2	1.29	0.5	
Faculty			
N_{3j}	150	50	200
n_{2j}	4	1	5
$\bar{y}_{3j}$	4	2	
s_{3j}^2	2.67	0	

For a survey on the athletic facilities on the campus, in a small sample of ten undergraduate students, $n_{11} = 5$ male and $n_{12} = 5$ females were observed. The number of hours a week of utilizing the facilities for the five male students were $y_{11j} = (3, 5, 3, 2, 4)$, and for the five female students, they were $y_{12j} = (3, 1, 2, 2, 1)$. Similarly, in a sample of five graduate students, $n_{21} = 3$ males and $n_{22} = 2$ females were observed. The number of hours for these groups were $y_{21j} = (4, 6, 4)$ and $y_{22j} = (3, 2)$. For a sample of five from the faculty, $n_{31} = 4$ and $n_{32} = 1$, with the number of hours $y_{31j} = (4, 4, 6, 2)$ and $y_{32j} = 2$. The sizes of these groups along with their sample means and variances are presented in Table 6.3 and Figure 6.1.

From the above observations, an estimate of the average for the undergraduate male students is $\bar{y}_{11} = 3.4$ hours. From (6.22), $v(\bar{y}_{11}) = [(2000 - 5)/2000](1.3/5) = 0.2594$, and hence S.E.$(\bar{y}_{11}) = 0.51$. Similarly, for the male graduate students, $\bar{y}_{21} = 4.67$, $v(\bar{y}_{21}) = [(600 - 3)/600](1.29/3) = 0.4279$ and S.E.$(\bar{y}_{21})$. For the male faculty, $\bar{y}_{31} = 4$, $v(\bar{y}_{31}) = [(150 - 4)/150](2.67/4) = 0.6497$ and S.E.$(\bar{y}_{31}) = 0.81$.

An unbiased estimator of the total $Y_{gj} = N_{gj}\bar{Y}_{gj}$ is $\hat{Y}_{gj} = N_{gj}\bar{y}_{gj}$. If N_{gj} is not known, it is replaced by $\hat{N}_{gj}$ and the procedure in Section 6.4 can be used to find its average variance.

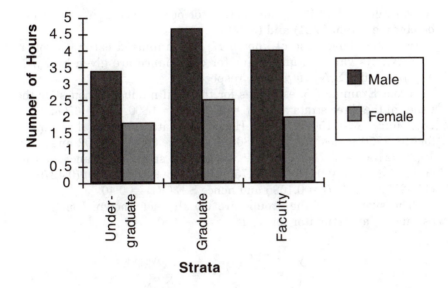

Figure 6.1. Usage of the athletic facility by the male and female groups.

6.8 Totals and means of subpopulations in strata

In the population, the jth group consists of $N_j = \Sigma_g^G N_{gj}$ units with total, mean, and variance:

$$Y_j = \sum_{g}^{G} Y_{gj}$$

$$\bar{Y}_j = Y_j/N_j,$$

and

$$S_j^2 = \sum_g \sum_i (y_{gij} - \bar{Y}_j)^2/(N_j - 1). \tag{6.23}$$

If the sizes N_{gj} of the jth group in the strata are known, an unbiased estimator of Y_j is

$$\hat{Y}_j = \sum_{g}^{G} \hat{Y}_{gj} = \sum_{g}^{G} N_{gj}\, \bar{y}_{gj}. \tag{6.24}$$

Its variance $\Sigma_g N_{gj}^2 \ V(\bar{y}_{gj})$ and estimator of variance $\Sigma_h N_{gj}^2 \ v(\bar{y}_{gj})$ can be obtained from (6.21) and (6.22).

For the above case of known N_{gj}, an unbiased estimator of $\overline{Y}_j$ is $\hat{Y}_j/N_j$. Its variance and estimator of variance are given by $\Sigma_g N_{gj}^2$ $V(\bar{y}_{gj})/N_j^2$ and $\Sigma_g N_{gj}^2 \ v(\bar{y}_{gj})/N_j^2$, respectively.

From Example 6.5, estimates for the total number of hours for the males in the three strata are $\hat{Y}_{11} = 2000(3.4) = 6800$, $\hat{Y}_{21} = 600(4.67) = 2802$, and $\hat{Y}_{31} = 150(4) = 600$. Hence, estimates of the total and mean for the 2750 males are $\hat{Y}_1 = 6800 + 2802 + 600 = 10,202$ and $\overline{\hat{Y}}_1 = 10,202/2750 = 3.71$ hours per week. The sample variance of this estimator is $v(\overline{\hat{Y}}_1) = (2000/2750)^2(0.2594) + (600/2750)^2(0.4279) + (150/2750)^2(0.6497) = 0.1595$ and hence S.E.$(\overline{\hat{Y}}_1) = 0.40$.

The sizes N_{gj} of the groups are usually not known. Using their estimates, an estimator for $\overline{Y}_j$ is

$$\overline{\hat{Y}}_j = \frac{\hat{Y}_j}{\hat{N}_j} = \frac{\sum_1^G \hat{N}_{gj} \bar{y}_{gj}}{\sum_1^G \hat{N}_{gj}} = \frac{\sum_1^G N_g \frac{n_{gj}}{n_g} \bar{y}_{gj}}{\sum_1^G N_g \frac{n_{gj}}{n_g}}. \tag{6.25}$$

This estimator resembles the ratio of two sample means and it is biased since the denominator is a random variable; see Chapter 9. For large samples, its variance is approximately given by

$$V(\overline{\hat{Y}}_j) = \frac{1}{N_j^2} \sum_1^G \frac{(N_g - n_g)N_{gj}}{n_g} S_{gj}^2. \tag{6.26}$$

An estimator of this variance can be obtained by replacing N_{gj}, N_j, and S_{gj}^2 by their unbiased estimators. Durbin (1958) and Cochran (1977) present an alternative estimator for (6.26).

An estimator for the difference $(\overline{Y}_j - \overline{Y}_{j'})$ of the means of two domains is $(\overline{\hat{Y}}_j - \overline{\hat{Y}}_{j'})$. The variance of this difference depends on the sample sizes n_g in the strata and the observed sizes n_{gj} and $n_{gj'}$ in the domains.

6.9   Stratification: Proportions and counts of subpopulations

Proportions and totals of qualitative characteristics for subpopulations in strata can be estimated through the notation in Section 6.6 and by suitably modifying the notation in Section 6.8. The following example provides an illustration.

Example 6.6. Off-campus sports: In the samples of Example 6.5, one each of the undergraduate and graduate male students and two of the male faculty participated in off-campus sports. Thus, $p_{11} = 1/5$, $p_{21} = 1/3$, and $p_{31} = 2/4$ are the estimates for the proportions in the three strata who participate in these sports. The sample variances of these estimates are $v(p_{11}) = [(2000 - 5)/2000 \times 4](1/5)(4/5) = 0.0399$, $v(p_{21}) = [(600 - 3)/600 \times 2](1/3)(2/3) = 0.1106$, and $v(p_{31}) = [(150 - 4)/150 \times 3](1/2)(1/2) = 0.0811$.

From these observations, an estimate for the proportion of the 2750 males participating in off-campus sports is $\hat{P}_1 = [2000(1/5) + 600(1/3) + 150(1/2)]/2750 = 0.2455$ or close to 25%. The sample variance of this estimator is $v(\hat{P}_1) = [2000^2(0.0399) + 600^2(0.1106) + 150^2(0.0811)]/(2750)^2 = 0.0266$ and hence S.E.$(\hat{P}_1) = 0.1632$.

Exercises

6.1. From the data in Table 6.1, (a) estimate the total number of beds in the second group and (b) find its S.E. when its size is known and when it is estimated.

6.2. (a) Using the figures in Table 6.1, estimate the total number of patients in the second group and find the S.E. of the estimate when the actual size is known. (b) Find the estimate and its S.E. when the actual size is estimated.

6.3. Using the sample data in Table 6.2 and the additional information in Example 6.4, estimate the following quantities and find their S.E.: (a) Proportions and numbers of residences with at least four persons, separately for the owner-occupied and rented households. (b) Differences of the above proportions and numbers. (c) Differences of the proportions of the single-member and two-member households in the rented households. In (a) and (b), estimate the numbers of households using the known sizes and by estimating the sizes.

6.4. With the sample data in Table 6.2 and Example 6.4, (a) estimate the proportion and total number of residences in that town with six or more persons and find the S.E. of the estimates. (b) Find the 90% confidence limits for this proportion and total.

6.5. Using the sample data in Table 6.2, estimate the average number of persons per household and find the 95% confidence limits for the average.

6.6. With the sample information presented in Table 6.3, estimate the average number of hours of utilizing the athletic facilities by the 2450 females and find the S.E. of the estimate.

6.7. With the information presented in Table 6.3 estimate the differences of the average number of hours for the 2750 males and the 2450 females, and find the S.E. of the estimate.

6.8. In the samples of Example 6.5, two of the undergraduate and one of the graduate female students but none of the female faculty participated in the off-campus sports. Estimate the proportion of the 2450 females participating in such sports and find the S.E. of the estimate.

6.9. *The New York Times* of February 25, 1994, summarized the results of a survey conducted by Klein Associates, Inc. on 2000 lawyers on sexual advances in the office. Between 85 and 98% responded to the questions in the survey; 49% of the responding women and 9% of the responding men agreed that some sorts of harassment exist in the offices. Assume that the population of lawyers is large and there are equal number of female and male lawyers, and ignore the nonresponse; that is, consider the respondents to be a random sample of the 2000 lawyers. (a) Find the standard errors for the above percentages. (b) Estimate the difference of the percentages for women and men, and find the S.E. of the estimate.

6.10. For the survey in Exercise 6.9, find the sample sizes needed for the following requirements: (a) The standard errors for women and men should not exceed 4 and 1%, respectively, and (b) the S.E. for the difference of the percentages for women and men should not exceed 3%.

6.11. Without the assumption in Exercise 6.9, find the standard errors for the women, men, and their difference.

6.12. *Project.* Consider the two groups of states in Table T5 in the Appendix with the total enrollments below 40,000 and 40,000 or more, and find their means and variances. For each of the 210 samples of size six from the ten states, find the means, variances, and the differences of the means. (a) For the first group, show that   for each observed sample size, the average of the sample means coincides with the actual mean of that group and their variance coincides with (6.7). (b) Show that the average of the variances in (a) over the sample sizes is larger than that obtained by replacing n_i by $n(N_i/N)$, as described in Section 6.3.

CHAPTER 7

Cluster Sampling

7.1   Introduction

A cluster is usually a large unit consisting of smaller units, also known as the **elements**. Cities, city blocks, colleges, dormitories, residence halls, shopping malls, and apartment complexes are examples of clusters. In agricultural, demographic, economic, and political surveys, the geographical area of interest is divided into clusters with counties, provinces, or enumeration districts as elements. For some surveys, proximity of the elements is used as a guide for cluster formation.

Interest in estimates for individual clusters, convenience of sampling, and reduction of traveling costs for interviews are some of the reasons for dividing a population into clusters. For a study on elementary schools in a large region, the area can be divided into a few clusters with each of the regions containing a certain number of schools. The needed information can be obtained by visiting the schools in a random sample of the clusters. A random sample of the schools from the entire region would be spread more evenly in the area, but the distances between the selected units may increase the costs of the survey. In such cases, costs should be weighed against precision for preferring cluster sampling.

To estimate the population totals, means, and proportions, it becomes convenient in some situations to consider clusters of **equal sizes**, each containing the same number of elements. To increase the precisions in the case of **unequal size** clusters, the estimators can be adjusted for their sizes through what is known as the ratio method and similar procedures. Alternatively, precisions of the estimators can also be increased by selecting the cluster units with probabilities related to their sizes, instead of selecting them through simple random sampling.

In large-scale multistage surveys, the population is first divided into strata, and samples from the clusters in each stratum are selected with equal or unequal probabilities.

7.2   Clusters of equal sizes

For the sake of illustration, data on the number of establishments and employment in 54 counties of New York State, divided into nine clusters each containing six counties are presented in Table 7.1.

When the population consists of N clusters with M elements in each, let y_{ij}, $i = 1, ..., N$ and $j = 1, ..., M$ denote the values of a characteristic of interest. The total and mean of the ith

Table 7.1.   Employment in 54 counties of New York State, divided into nine clusters of size six each; number of establishments (1000s) and number of persons employed (1000s).

	Establishments	Persons Employed	Establishments	Persons Employed	Establishments	Persons Employed
		1		4		7
	4.2	63.8	0.7	11.7	1.0	13.7
	21.3	341.2	3.2	41.9	1.1	15.1
	0.6	6.1	1.7	22.2	1.2	13.8
	1.2	14.4	0.8	10.1	0.5	4.1
	0.7	7.8	1.8	25.9	3.1	52.5
	1.0	11.6	1.9	29.0	8.0	132.0
Total	29.0	444.9	10.1	140.8	14.9	231.2
Mean	4.8	74.2	1.7	23.5	2.5	38.5
		2		5		8
	15.1	341.0	2.0	20.7	2.6	34.2
	2.0	25.0	1.7	21.6	4.5	84.7
	1.4	15.8	2.0	24.2	1.1	11.7
	0.4	4.1	0.5	3.8	1.1	8.4
	0.5	8.6	0.9	9.1	1.3	12.3
	1.5	17.5	1.6	17.8	1.9	17.7
Total	20.9	396.0	8.7	105.9	12.5	169.0
Mean	3.5	66.0	1.5	17.7	2.1	28.2
		3		6		9
	0.3	3.1	1.1	14.5	3.6	41.8
	1.9	34.9	1.3	13.7	5.5	90.0
	0.9	14.0	1.1	10.3	6.2	69.9
	11.2	203.9	1.9	21.3	1.7	12.4
	5.1	74.9	1.0	10.2	6.8	73.4
	1.2	11.5	2.9	29.7	13.1	170.5
Total	20.6	342.3	9.3	99.7	36.9	458.0
Mean	3.4	57.1	1.6	16.6	6.2	76.3

Source: U.S. Bureau of the Census; 1985 figures.

Table 7.2. Means and variances of the nine clusters of Table 7.1.

| | No. of Establishments | | No. of Persons Employed | |
Cluster	Mean	Variance	Mean	Variance
1	4.83	66.89	74.15	17,587.70
2	3.48	32.77	68.67	17,852.40
3	3.43	17.33	57.05	5,842.58
4	1.68	0.82	23.46	139.01
5	1.45	0.38	16.03	67.03
6	1.55	0.54	16.62	57.43
7	2.48	8.10	38.53	2,378.86
8	2.08	1.73	28.17	851.12
9	6.15	15.09	76.33	2,876.83

cluster are

$$y_i = \sum_{j=1}^{M} y_{ij} = y_{i1} + y_{i2} + \cdots + y_{iM} \qquad (7.1)$$

and

$$\bar{y}_i = \frac{y_i}{M}. \qquad (7.2)$$

The totals and means for the above nine clusters of New York State are presented in Table 7.1. The means of the clusters, along with their variances are presented in Table 7.2.

The population **mean per unit** is obtained by summing the N totals in (7.1) and dividing by N; that is,

$$\bar{Y} = \frac{1}{N} \sum_{1}^{N} y_i. \qquad (7.3)$$

The **mean per element** is obtained by summing the observations of the NM elements and dividing by NM; that is,

$$\bar{\bar{Y}} = \frac{1}{M_o} \sum_{1}^{N} \sum_{1}^{M} y_{ij} = \frac{\bar{Y}}{M}, \qquad (7.4)$$

where $M_o = NM$ is the total number of elements.

For the above illustration, $\bar{Y}$ can denote the average number of establishments per cluster and $\bar{\bar{Y}}$ the average number of the establishments per county. Similarly, $\bar{Y}$ can denote the employment per cluster and $\bar{\bar{Y}}$ the employment per county. For another illustration, if an agricultural region, for example, is divided into N fields with M farms in each, $\bar{Y}$ can denote the production per field and $\bar{\bar{Y}}$ the production per farm. Both types of means are important in practice.

The population variance is

$$S^2 = \frac{\sum_1^N \sum_1^M (y_{ij} - \bar{\bar{Y}})^2}{NM - 1}. \tag{7.5}$$

The variance of the ith cluster is $s_i^2 = \sum_1^M (y_{ij} - \bar{y}_i)^2/(M - 1)$. As shown in Appendix A7, (7.5) can be expressed as

$$S^2 = \frac{N(M - 1)}{NM - 1} S_w^2 + \frac{N - 1}{NM - 1} S_b^2 \tag{7.6}$$

$$= \frac{(M - 1)S_w^2 + S_b^2}{M}, \tag{7.6a}$$

where $S_w^2 = (\sum_1^N s_i^2)/N$ and $S_b^2 = M\sum (\bar{y}_i - \bar{\bar{Y}})^2/(N - 1)$ are the **within MS** (mean square) and **between MS** with $N(M - 1)$ and $(N - 1)$ d.f., respectively. The expression in (7.6a) is an approximation of (7.6) for large N.

For the 54 counties of Table 7.1, the average $\bar{\bar{Y}}$ and variance S^2 for the number of employed persons are 44.34 and 5057.79. Corresponding figures for the number of establishments are 3.02 and 15.94, respectively. The correlation between the number of establishments and the number of persons employed is 0.9731.

7.3   Estimation of the means

The population means per unit and per element can be estimated by selecting a sample of size n from the N clusters, and recording the M observations in each of the selected clusters. The sample means on the unit and element basis are

$$\bar{y} = \frac{1}{n} \sum_1^n y_i \tag{7.7}$$

and

$$\bar{\bar{y}} = \frac{1}{nM} \sum_1^n \sum_1^M y_{ij} = \frac{\bar{y}}{M}. \tag{7.8}$$

These means are unbiased for $\bar{Y}$ and $\bar{\bar{Y}}$, respectively. Their variances are

$$V(\bar{y}) = \frac{(1-f)}{n} \frac{\sum_1^N (y_i - \bar{Y})^2}{N-1} \tag{7.9}$$

and

$$V(\bar{\bar{y}}) = \frac{1}{M^2} V(\bar{y})$$

$$= \frac{(1-f)}{nM^2} \frac{\sum (y_i - \bar{Y})^2}{N-1} = \frac{(1-f)}{n} \frac{\sum (\bar{y}_i - \bar{\bar{Y}})^2}{N-1}. \tag{7.10}$$

From the definition of S_b^2 and from (7.9), this variance can also be expressed as

$$V(\bar{y}) = \frac{(1-f)S_b^2}{nM} \tag{7.11}$$

Observe from (7.10) that the variances of $\bar{y}$ and $\bar{\bar{y}}$ will be small if the cluster totals or means do not vary much.

Example 7.1. Estimating employment: Table 7.1 contains the data for private nonfarm establishments. This type of establishment includes manufacturing, retail trade, finance, insurance, and real estate firms and services. One of the counties with only 200 establishments is combined with a neighboring county. New York City alone had 103 thousand establishments with 1988 thousand persons employed, and the adjacent six counties had 177 thousand establishments with a total of 2122 thousand employed persons. Figures for these seven counties are not included in Table 7.1.

If one selects a random sample, for example, of size $n = 3$ from the nine clusters of Table 7.1, any one of the $_9C_3 = 84$ combinations of the clusters may be selected.

For the employment figures, from Table 7.1, $\Sigma_1^N(\bar{y}_i - \bar{\bar{Y}})^2/(N-1)$ equals 620.8. If $n = 3$, from (7.10), $V(\bar{y}) = 137.96$. Similarly, for the establishments, $\Sigma_1^N(\bar{y}_i - \bar{\bar{Y}})^2/(N-1)$ equals 2.63 and $V(\bar{y}) = 0.58$ if $n = 3$.

7.4 Comparison with simple random sampling

As mentioned in Section 7.1, a simple random sample of size nM drawn without replacement from the NM elements is spread more evenly in the population than a sample of n clusters. The mean of these nM elements is unbiased for $\bar{\bar{Y}}$ and has variance:

$$V(\bar{y}_{srs}) = \frac{(1-f)}{nM}S^2. \qquad (7.12)$$

Before preferring cluster sampling over simple random sampling, their relative precisions may be compared. By substituting (7.6) in (7.12) and dividing by (7.11), precision of cluster sampling relative to simple random sampling is given by

$$\frac{V(\bar{y}_{srs})}{V(\bar{y})} = \frac{S^2}{S_b^2} = \frac{N(M-1)}{(NM-1)F} + \frac{N-1}{NM-1}, \qquad (7.13)$$

where $F = S_b^2/S_w^2$ approximately follows the F-distribution with $(N-1)$ and $N(M-1)$ d.f., provided N and M are large. Percentiles of this distribution are available in the standard statistical tables. For large N and M, the expression in (7.13) can also be expressed as $[(M-1)/F + 1]/M$.

Equation 7.13 demonstrates that cluster sampling has higher precision than simple random sampling if S_w^2 is large and S_b^2 is small. This requirement is just the opposite of what is needed for high precision with stratification. Note also that the sample size has no effect on the relative precision of these two procedures of sampling. Following Kish (1965), the ratio $V(\bar{y})/V(\bar{y}_{srs})$ is known as the design effect, **deff**. The magnitude of this ratio enables one to examine the different factors affecting $V(\bar{y})$.

As has been seen, in the case of stratification, samples are selected from each of the strata. For cluster sampling, a sample of clusters is selected and observations on all the elements of the selected clusters are recorded.

Example 7.2. Gains relative to simple random sampling: Since S^2 for the employment figures in Example 7.1 is 5057.79, the variance of $\bar{y}_{srs}$ from (7.12) for a sample of $nM = 18$ counties from the $NM = 54$ counties is $[(9 - 3)/(9 \times 3)](5057.79) = 187.33$. This variance is larger than the variance of 137.96 for $\bar{y}$ found in Example 7.1. The gain in precision for cluster sampling is $(187.33 - 137.96)/137.6 = 0.36$, that is, 36%.

The estimator for S_b^2 is presented in Section 7.5 and for S^2 in Appendix A7. The relative precision in (7.13) can be found from these estimators.

7.5   Estimation of the standard error

From a sample of n clusters, unbiased estimators of the variances in (7.9) and (7.10) are

$$v(\bar{y}) = \frac{(1-f)}{n} \frac{\sum^n (y_i - \bar{y})^2}{n-1} \tag{7.14}$$

and

$$v(\bar{\bar{y}}) = \frac{(1-f)}{n} \frac{\sum^n (\bar{y}_i - \bar{\bar{y}})^2}{n-1} = \frac{(1-f)}{n} \frac{s_b^2}{M} \tag{7.15}$$

where $s_b^2 = [M\sum^n(\bar{y}_i - \bar{\bar{y}})^2]/(n-1)$ is the sample between MS with $(n-1)$ d.f.

Example 7.3. Standard error for estimating the employment: If clusters 1, 7, and 9 appear in a sample of three from the nine clusters of Table 7.1, the means for the employments are 74.15, 38.53, and 76.33. Hence, an estimate for the average employment per county is $\bar{\bar{y}} = (74.15 + 38.53 + 76.33)/3 = 63$ (thousands). Since $\sum_1^n(\bar{y}_i - \bar{\bar{y}})^2/(n-1)$ is 450.4, $v(\bar{\bar{y}}) = [(9 - 3)/(9 \times 3)](450.4) = 100.1$. Note that $s_b^2 = 6(450.4) = 2702.4$, and the same value is obtained from the right-hand side expression of (7.15).

The actual mean is $\bar{\bar{Y}} = 44.34$ (thousands). As seen in Example 7.1, the actual variance of $\bar{y}$ equals 137.96, and $S_b^2 = 6(620.8) = 3724.8$.

7.6   Optimum cluster and sample sizes

As seen in Sections 7.3 and 7.4, larger values of S_w^2 and smaller values of S_b^2 result in higher precision for cluster sampling. This result can be used for dividing a population into clusters of suitable sizes. For several agricultural surveys, S_w^2 was found to be proportional to M^h, where h is usually between zero and 2. This observation may also be valid for certain types of demographic and economic surveys. Such information can be used to find the optimum sizes for M and n, for example, by prescribing an upper limit to the variance in (7.11).

Since a sample of n clusters are selected and observations are made on the M elements in each of the selected clusters, the cost of obtaining information from the sample can be a multiple of Mn. In some studies, travel costs are also considered; see Cochran (1977; Chap. 9), for example. The optimum values of M and n are obtained by minimizing $V(\bar{y})$ for the available budget, or minimizing the cost for a required precision. The above type of information regarding S^2 and S_w^2 can be used for this purpose. For telephone and mail surveys, travel costs may not be applicable, and the relevant cost functions should be chosen.

7.7   Clusters of unequal size

Most of the natural clusters of the type described in Section 7.1 usually tend to be of unequal sizes. When the population consists of N clusters with M_i elements in the ith cluster, let y_{ij}, $i = 1, ..., N$ and $j = 1, ..., M_i$, denote the observations. The cluster totals are $y_i = \sum_1^{M_i} y_{ij}$. As before, the means per unit and per element, respectively, are

$$\bar{Y} = \frac{\sum_1^N y_i}{N} \quad \text{and} \quad \bar{\bar{Y}} = \frac{\sum_1^N y_i}{M_o} = \frac{\bar{Y}}{\bar{M}}, \quad (7.16)$$

where $M_o = \sum_1^N M_i$ is the total number of elements and $\bar{M} = M_o/N$ is the average number of elements per cluster.

One type of grouping of the neighboring counties of New York State, presented in Table 7.1, resulted in nine clusters of unequal size. Their sizes M_i, and the totals x_i and y_i for the number of establishments and the number of employed persons along with their standard deviations and correlations are presented in Table 7.3.

For a sample of n clusters, the estimator $\bar{y}$ for $\bar{Y}$ and its variance $V(\bar{y})$ have the same forms as (7.7) and (7.9), respectively. The unbiased estimator for $V(\bar{y})$ takes the same form as (7.14).

An unbiased estimator of $\bar{\bar{Y}}$ is given by $\hat{\bar{\bar{Y}}} = \bar{y}/\bar{M}$, which has the variance of $V(\bar{y})/\bar{M}^2$.

Table 7.3. Unequal size clusters: totals and variances.

Cluster i	Cluster Size M_i	Number of Establishments		Number Employed	
		Total	Variance	Total	Variance
1	10	49.8	52.04	860.1	18,398.01
2	8	20.5	12.52	322.4	4,427.36
3	3	4.1	0.85	67.0	286.14
4	4	7.1	3.32	124.1	1,281.49
5	4	6.5	0.27	70.8	47.81
6	5	6.4	0.132	70.6	20.66
7	5	9.1	3.45	118.5	837.23
8	7	19.5	6.27	273.2	1,968.01
9	8	39.9	15.46	487.4	2,871.45
Total	54	162.9		2,394.1	
Mean	6	18.1		266.01	
S.D.	2.35	16.42		265.28	

Correlations of employment totals with establishment totals and cluster sizes are 0.98 and 0.92, respectively. Correlation of establishment totals with cluster sizes is 0.92.

Example 7.4. Average employment: Consider the sample size $n = 3$ for the nine clusters in Table 7.3. In this case, from (7.9), $V(\bar{y}) = (9 - 3)$ $(265.28)^2/(9 \times 3) = 15,638.55$. Since $\bar{M} = 6$, $V(\bar{\bar{y}}) = 15,653.55/36 = 434.4$.

As in the case of equal cluster sizes, the above estimators will have small variances if the cluster totals y_i do not vary much.

7.8   Alternative estimation with unequal sizes

If the sizes are unequal, $\bar{\bar{Y}}$ can also be estimated from

$$\hat{\bar{\bar{Y}}}_R = \frac{\sum_1^n y_i}{\sum_1^n M_i} = \frac{\bar{y}}{\bar{m}}, \tag{7.17}$$

where $\bar{m} = \sum_1^n M_i/n$ and the subscript R denotes the ratio.

Although $\bar{y}$ and $\bar{m}$ are unbiased for $\bar{Y}$ and $\bar{M}$, the ratio estimator in (7.17) is not unbiased for $\bar{\bar{Y}} = \bar{Y}/\bar{M}$. For large n, as shown in Appendix A7, the bias of this estimator becomes small and its variance is approximately given by

$$V(\hat{\bar{Y}}_R) = \frac{(1-f)}{n\overline{M}^2}(S_y^2 + \overline{Y}^2 S_m^2 - 2\overline{Y}\rho S_y S_m). \tag{7.18}$$

In this expression, S_y^2 and S_m^2 are the variances of (y_i, M_i) of the N clusters, and ρ is the correlation of y_i and M_i. Note that the covariance of y_i with M_i is $S_{ym} = \rho S_y S_m$.

As noted at the end of Section 7.7, for simple random sampling, the variance of $\hat{\bar{Y}} = \bar{y}/\overline{M}$ is given by $V(\bar{y})/\overline{M}^2$. The variance of $\overline{Y}_R$ in (7.18) will be smaller than that of $\hat{\bar{Y}}$ if $\rho \geq C_m/2C_y$, where C_y and C_m are the coefficients of variations of y_i and M_i, respectively.

Example 7.5. Ratio estimation adjusting for the sizes: From Table 7.3, the ratio of the totals $\Sigma_1^N y_i$ to $\Sigma_1^N M_i$ is $(2394.1/54) = 44.34$, which is the same as $\hat{\bar{Y}} = 266.01/6$. For a sample of size three from the nine clusters, the variance of $\overline{Y}_R$ in (7.18) is

$$V(\hat{\bar{Y}}_R) = \frac{9-3}{9(3)(36)}[(265.28)^2 + (44.34)^2(2.35)^2$$
$$- 2(44.34)(0.92)(265.28)(2.35)] = 187.47.$$

This variance is only about two fifths of the variance of 434.4 found for the sample mean in Example 7.4.

As can be seen from the comparison of $V(\hat{\bar{Y}}_R)$ in (7.18) with $V(\hat{\bar{Y}}) = V(\bar{y})/\overline{M}^2$, high positive correlation between the employment and the cluster size has considerably reduced the variance of $\overline{Y}_R$ relative to that of $\hat{\bar{Y}}$.

An estimator for (7.18) is given by

$$V(\hat{\bar{Y}}_R) = \frac{(1-f)}{n\overline{M}^2}(S_y^2 + \hat{\bar{Y}}_R^2 s_m^2 - 2\hat{\bar{Y}}_R s_{ym}), \tag{7.19}$$

where s_{ym} is the sample covariance of y_i and M_i.

For $\hat{\bar{Y}}$, as an alternative to the sample mean $\bar{y}$, one may consider the *ratio estimator*:

$$\hat{\bar{Y}}_R = \frac{\sum_1^n y_i}{\sum_1^n M_i} \overline{M}. \tag{7.20}$$

Following the procedure in Appendix A7, the bias of this estimator becomes small as n increases. The approximate variance of (7.20) and its estimator are given by

$$V(\hat{\bar{Y}}_R) = \frac{(1-f)}{n}(S_y^2 + \bar{\bar{Y}}^2 S_m^2 - 2\bar{\bar{Y}}\rho S_y S_m) \qquad (7.21)$$

and

$$v(\hat{\bar{Y}}_R) = \frac{(1-f)}{n}(s_y^2 + \hat{\bar{Y}}_R^2 s_m^2 - 2\hat{\bar{Y}}_R s_{ym}). \qquad (7.22)$$

Note that $V(\hat{\bar{Y}}_R) = \bar{M}^2 V(\hat{\bar{Y}}_R)$. Thus, for the above example, $V(\hat{\bar{Y}}_R) = 36(187.47) = 6748.92$. This variance is only about two fifths of the variance of 15,638.55 for $\bar{y}$ found in Example 7.4.

7.9 Proportions and percentages

When a population is divided into clusters, estimation of the total numbers and proportions or percentages related to qualitative characteristics also is usually of interest.

Let c_i and $p_i = (c_i/M_i)$, $i = 1, 2,...,N$, denote the number and proportion of the M_i elements of the ith cluster having an attribute of interest.

For each of the nine clusters described in Sections 7.7 and 7.8, the numbers c_i and proportions p_i of counties that have more than 1000 establishments are presented in columns 3 and 4 of Table 7.4. The numbers and proportions of the counties that have more than 10,000 employees are presented in columns 5 and 6.

Table 7.4. Composition of the 54 counties.

| Cluster | No. of Counties | No. and Proportion of Counties with More Than | | | |
| | | 1000 establishments | | 10,000 employees | |
i	M_i	c_i	p_i	c_i	p_i
1	10	7	0.70	8	0.80
2	8	6	0.75	6	0.75
3	3	2	0.67	2	0.67
4	4	2	0.50	4	1.00
5	4	3	0.75	4	1.00
6	5	5	1.00	5	1.00
7	5	4	0.80	4	0.80
8	7	6	0.86	5	0.71
9	8	8	1.00	8	1.00
Means	6		4.78	0.78	5.11	0.86

Average of the proportions

The average of the N proportions is

$$\bar{P} = \frac{1}{N}\sum_{1}^{N}p_i.$$ (7.23)

For either of the two characteristics in Table 7.4, $\bar{P}$ is the average of the nine proportions. For a sample of n clusters, an unbiased estimator of $\bar{P}$ is

$$\hat{\bar{P}} = \frac{1}{n}\sum_{1}^{n}p_i.$$ (7.24)

Its variance and estimator of variance are obtained from $V(\bar{y})$ and $v(\bar{y})$ in (2.12) and (2.14) by replacing y_i with p_i.

> **Example 7.6.** Proportions of the counties for the establishments: As shown in Table 7.4, the average of the proportions of counties with more than 1000 establishments is 0.78. If this average is not known, one may estimate it from a random sample from the nine clusters. The variance of the nine proportions is 0.0227. For a sample of three clusters, following (2.12), $V(\hat{\bar{P}}) = (9 - 3)(0.0227)/(9 \times 3) = 0.005$ and S.E.$(\hat{\bar{P}}) = 0.07$.

Proportion of the totals or means

A total of $\sum_{1}^{N}c_i$ of the $M_o = \sum_{1}^{N}M_i$ elements have the attribute of interest. Their proportion is

$$P = \frac{\sum_{1}^{N}c_i}{M_o} = \frac{\sum_{1}^{N}M_ip_i}{M_o} = \frac{(\sum_{1}^{N}c_i)/N}{\bar{M}}.$$ (7.25)

In contrast to $\bar{P}$ in (7.23), this is the overall proportion of all the M_0 elements having the attribute of interest. If all the M_i are equal, it coincides with $\bar{P}$.

From a sample of n clusters, an unbiased estimator of P is

$$p = \frac{(\sum_{1}^{n}c_i)/n}{\bar{M}} = \frac{\bar{c}}{\bar{M}}$$ (7.26)

and its variance is given by $V(\bar{c})/\overline{M}^2$. Note that $V(\bar{c})$ and its estimate can be obtained from (2.12) and (2.14) by replacing y_i with c_i.

Another estimator for P is

$$\hat{P} = \frac{\sum_1^n c_i}{\sum_1^n M_i} = \frac{\sum_1^n M_i p_i}{\sum_1^n M_i}, \tag{7.27}$$

which is of the ratio type. Its large sample variance is given by

$$V(\hat{P}) = \frac{(1-f)}{n\overline{M}^2} \frac{\sum_1^N (c_i - PM_i)^2}{N-1}. \tag{7.28}$$

An estimator of this variance is obtained from

$$v(\hat{P}) = \frac{(1-f)}{n\overline{M}^2} \frac{\sum_1^N (c_i - \hat{P}M_i)^2}{n-1}. \tag{7.29}$$

If c_i is highly correlated with M_i, $\hat{P}$ will have smaller variance than p. If $\overline{M}$ is not known, for the variance in (7.28), replace it with the sample mean $\Sigma M_i/n$ of the sizes.

Example 7.7. Establishments and the gain in precision for the ratio estimation: For the number of counties of Table 7.4 that have more than 1000 establishments, $P = (4.78/6) = 0.7967$. The variance of c_i is 4.67 and for a sample of three clusters $V(\bar{c}) = (9 - 3)(4.67)/(9 \times 3) = 1.0378$. To estimate this proportion with a sample of three clusters, $V(p) = 1.0378/36 = 0.0288$ and hence S.E.$(p) = 0.17$. For the alternative estimator in (7.27), from (7.28), $V(\hat{P}) = 0.005$ and S.E.$(\hat{P}) = 0.07$. The relative gain in precision for the ratio estimation is $(0.029 - 0.005)/0.005 = 4.8$, that is, 480%, which is substantial.

7.10   Stratification

In several large-scale surveys, the population is first divided into strata, with each stratum consisting of a certain number of clusters. For example, the nine clusters of Table 7.3 can be divided into three strata containing the clusters (1, 2, 8, 9), (4, 7), and (3, 5, 6).

With $\hat{Y}_g$ denoting the estimate for the total of the gth stratum obtained through a sample, an estimator for the population total is $\hat{Y} = \Sigma \hat{Y}_g$.

The variance of $\hat{Y}$ and its estimate are given by $\Sigma V(\hat{Y}_g)$ and $\Sigma v(\hat{Y}_g)$. The following example examines the total and average of the employment.

Example 7.8. Total and average of the employment: For the above type of stratification, from Table 7.3, the standard deviations of the employment figures in the three strata equal 265.83, 3.96, and 2.14. If a sample of one cluster is drawn randomly from each of these strata, an unbiased estimator of the total takes the form $\hat{Y} = \Sigma_1^3 N_g y_g$. The variance of this estimator is $V(\hat{Y}) = \Sigma N_g(N_g - 1)S_g^2$.

An unbiased estimator of $\bar{\bar{Y}}$ is $\hat{\bar{Y}} = \hat{Y}/M_o$, which has the variance $V(\hat{\bar{Y}}) = \Sigma N_g(N_g - 1)S_g^2/M_0^2 = [4(3)(265.83)^2 + 2(1)(3.96)^2 + 3(2)(2.14)^2]/(54)^2 = 290.32$. This variance as expected is smaller than the variance of 434.4 for the sample mean found in Example 7.4.

7.11 Unequal probability selection

Chapters 2 to 6 and the preceding sections of this chapter have considered selection of the units through simple random sampling, that is, with equal probabilities and without replacement. In some applications, the units are selected with unequal probabilities, and unbiased estimators for the population totals and means are obtained. If these probabilities are related to the characteristics of interest, the resulting estimators can have smaller variances than those obtained through simple random sampling.

In cluster sampling, the probabilities for selecting the units can depend on their sizes, the number of elements (M_i). They can also be based on one or more concomitant or supplementary variables highly correlated with the characteristics of interest.

A method of selecting the cluster units

For a simple method of selection with unequal probabilities, consider the nine cluster units in Table 7.3. Their cumulative sizes are 10, 18, 21, 25, 29, 34, 39, 46, and 54. The ranges associated with these figures are 1 to 10, 11 to 18, 19 to 21,...,47 to 54. One number is selected randomly between 1 and 54. If this number is 16, for example, the second cluster is selected into the sample. Since the above ranges correspond to the sizes of the clusters, the ith cluster is selected with probability $u_i = (M_i/M_o)$. If all the M_i are the same, $u_i = (1/N)$ and this procedure is the same as simple random sampling.

Continuing with the above procedure, if the ith unit appears at the first selection, it is removed from the list and the second unit is drawn with probability $M_j/(M_0 - M_i) = u_j/(1 - u_i)$. The next section examines the probabilities for selecting the units into a sample of size $n = 2$ using the above procedure.

Probabilities of selecting the units

The probabilities of selecting the jth unit after the ith unit and the ith unit after the jth unit, respectively, are

$$P(j|i) = \frac{u_j}{1 - u_i} \quad \text{and} \quad P(i|j) = \frac{u_i}{1 - u_j}. \tag{7.30}$$

The probability of selecting the ith unit into the sample, at the first or the second draw, is

$$\phi_i = u_i + \sum_{j \neq i}^{N} u_j \frac{u_i}{1 - u_j} = u_i\left(1 + T - \frac{u_i}{1 - u_i}\right) \tag{7.31}$$

where $T = \Sigma_1^N [u_i/(1 - u_i)]$.

The probability that the ith and jth units are drawn into the sample is

$$\phi_{ij} = u_i \frac{u_j}{1 - u_i} + u_j \frac{u_i}{1 - u_j} = \frac{u_i u_j(2 - u_i - u_j)}{(1 - u_i)(1 - u_j)}. \tag{7.32}$$

The above procedure can be continued for selecting more than two units into the sample. If the ith and the jth units appear at the first two draws, they are removed from the population and the third unit is drawn with probability $u_k/(1 - u_i - u_j)$. Continuation of this selection procedure provides a sample of n units.

The expression in (7.31) is the probability that the ith unit appears in the sample—the *inclusion probability*. As can be seen, ϕ_i is not exactly proportional to u_i. For the sake of illustration, consider four units with sizes $M_i = (15, 12, 10, 3)$ and employments $y_i = (105, 80, 45, 10)$ in thousands. The relative sizes of these units are $u_i = (0.375, 0.30, 0.25, 0.075)$. If two units are selected into the sample as above with these probabilities, from (7.31), $\phi_i = (0.6911, 0.6042, 0.5275, 0.1772)$. These inclusion probabilities are in the relative ratios $(0.346, 0.302, 0.264, 0.089)$, which differ slightly from the u_i.

If a sample of n units is selected randomly without replacement from the N population units, as seen in Chapter 2, the inclusion probability for the ith unit is $\phi_i = n/N = n(1/N)$, which is the same for all N units. Similarly, when $n = 2$ units are selected with unequal probabilities and without replacement, ϕ_i in (7.31) should equal $2u_i = (0.75, 0.6, 0.5, 0.15)$. The above inclusion probabilities $\phi_i = (0.6911, 0.6042, 0.5275, 0.1772)$ are close to these figures but not identical.

To find the initial probabilities p_i to make ϕ_i equal $2u_i$, Narain (1951) and Yates and Grundy (1953) replace u_i on the right-hand side of (7.31) with p_i and solve for the ϕ_i. They also establish the following general results for the selection of a sample of size n with unequal probabilities and *without* replacement.

$$\sum_i^N \phi_i = n, \tag{7.33a}$$

$$\sum_{j \neq i}^N \phi_{ij} = (n - 1)\phi_i, \qquad \sum_i^N \sum_{j \neq i}^N \phi_{ij} = n(n - 1), \tag{7.33b}$$

and

$$\sum_{j \neq i}^N (\phi_i \phi_j - \phi_{ij}) = \phi_i(1 - \phi_i). \tag{7.33c}$$

Note from (7.33a) that $\sum_{j \neq i}^N \phi_j = n - \phi_i$. With this result, (7.33c) follows from (7.33b).

Brewer (1963), Durbin (1967), and others have suggested procedures for selecting a sample of two units with probabilities proportional to u_i without replacement. Samford (1967) extends Brewer's procedure to the selection of more than two units.

7.12   Horvitz-Thompson estimator

For the estimation of the population total Y by selecting units with probabilities ϕ_i and without replacement, Horvitz and Thompson (1952) consider

$$\hat{Y}_{\text{HT}} = \sum_1^n \frac{y_i}{\phi_i}. \tag{7.34}$$

As shown in Appendix A7, this estimator is unbiased for Y.

Variance of the estimator

Yates and Grundy (1953) and Sen (1953) derive the variance of $\hat{Y}_{HT}$ and the two estimators for this variance presented below. As described in Appendix A7,

$$V(\hat{Y}_{HT}) = E(\hat{Y}_{HT}^2) - Y^2$$

$$= \sum_i^N \frac{1 - \phi_i}{\phi_i} y_i^2 + \sum_{i \neq j}^N \sum^N \frac{\phi_{ij} - \phi_i \phi_j}{\phi_i \phi_j} y_i y_j. \tag{7.35}$$

By substituting (7.33c) in the first term of (7.35), the variance can also be expressed as

$$V(\hat{Y}_{HT}) = \sum_i^N \sum_{j \neq i}^N \frac{\phi_i \phi_j - \phi_{ij}}{\phi_i^2} y_i^2 - \sum_{i \neq j}^N \sum^N \frac{\phi_i \phi_j - \phi_{ij}}{\phi_{ij}} y_i y_j$$

$$= \sum_i^N \sum_{i < j}^N (\phi_i \phi_j - \phi_{ij}) \left(\frac{y_i}{\phi_i} - \frac{y_j}{\phi_j} \right)^2. \tag{7.36}$$

For the illustration in Section 7.11 with the four units, the joint probabilities ϕ_{ij} and $(y_i/\phi_i) - (y_j/\phi_j)$ are presented in Table 7.5. From these figures, from (7.36), $V(\hat{Y}_{HT}) = 445.67$. For these units, $S_y^2 = 1716.67$, and for a simple random sample of $n = 2$ units, $V(\hat{Y}) = 4(4 - 2)$ $(1716.67)/2 = 6866.67$. Thus, $V(\hat{Y}_{HT})$ is only 6.5% of $V(\hat{Y})$.

As can be seen from the expressions in (7.34) or (7.36), the variance of the Horvitz–Thompson estimator vanishes when ϕ_i are proportional to y_i. Thus, selecting the population units with probabilities u_i and

Table 7.5. Probabilities of selection.

Clusters (i, j)	ϕ_{ij}	$(y_i/\phi_i) - (y_j/\phi_j)$
1, 2	0.3407	19.53
1, 3	0.1547	66.62
1, 4	0.0754	39.57
2, 3	0.2071	47.10
2, 4	0.0565	20.05
3, 4	0.0453	−27.05

making ϕ_i equal nu_i results in a small variance for $\hat{Y}_{HT}$, provided y_i are proportional to u_i. For cluster sampling, as noted earlier, the u_i can be based on the sizes M_i or the values of a supplementary variable.

The estimator $\hat{Y}_{HT}$ is of a general type obtained by sampling the population units with specified probabilities and without replacement. As seen in Chapter 2, for simple random sampling, all the ϕ_i are equal to (n/N) and $\hat{Y} = N\bar{y}$.

Variance estimators

From (7.35), an unbiased estimator for $V(\hat{Y}_{HT})$ is

$$v_1(\hat{Y}_{HT}) = \sum_i^n \frac{1 - \phi_i}{\phi_i^2} y_i^2 + \sum_{i \neq j}^n \sum^n \frac{\phi_{ij} - \phi_i\phi_j}{\phi_i\phi_j\phi_{ij}} y_i y_j. \tag{7.37}$$

As suggested by Yates and Grundy (1953) and Sen (1953), from (7.36), an alternative unbiased estimator is given by

$$v_2(\hat{Y}_{HT}) = \sum_i^n \sum_{i<j}^n \frac{\phi_i\phi_j - \phi_{ij}}{\phi_{ij}} \left(\frac{y_i}{\phi_i} - \frac{y_j}{\phi_j}\right)^2. \tag{7.38}$$

Both these estimators may take negative values.

Estimators for the mean per unit and the population mean

The estimator for the mean per unit $\bar{Y}$ in (7.3) is given by $\hat{\bar{Y}}_{HT} = \hat{Y}_{HT}/N$. Its variance is given by dividing (7.35) or (7.36) by N^2. Similarly, its variance estimator is obtained by dividing (7.37) or (7.38) by N^2. For the above illustration, $V(\hat{\bar{Y}}_{HT}) = 445.67/16 = 27.85$. For a simple random sample of two units, $V(\bar{y}) = 6866.67/16 = 429.17$.

The estimator for the population mean $\bar{\bar{Y}}$ in (7.4) is given by $\hat{\bar{\bar{Y}}}_{HT} = \hat{\bar{Y}}_{HT}/M_0$. Its variance is obtained by dividing (7.36) by M_0^2. Similarly, its estimator of variance is obtained by dividing (7.38) by M_0^2. Since $M_0 = 40$ for the above illustration, $V(\hat{\bar{\bar{Y}}}_{HT}) = 445.67/1600 = 0.2785$. For a simple random sample, $\hat{Y}/M_0$ is unbiased for $\bar{\bar{Y}}$, and its variance is given by $V(\hat{Y})/M_0^2$. For a sample of size two, the variance of this estimator is $6866.67/1600 = 4.29$.

7.13   Alternative approaches

Procedures for selecting two or more units with unequal probabilities and without replacement were suggested by Murthy (1957, 1967), J.N.K. Rao (1965), and others. Cochran (1977, pp. 258–270) describes these approaches in detail. Bayless and J.N.K. Rao (1970) empirically compare some of these procedures. Fellegi (1963) considers selection probabilities for rotation sampling. Cochran (1942) describes some of the estimation procedures with cluster units of unequal size. Brewer and Hanif (1983) summarize several methods for selecting units with unequal probabilities.

The following approaches can also be considered to obtain unbiased estimators for the population mean or total, and under suitable conditions they can result in small variances for the estimators.

Probabilities proportional to sum of sizes (PPSS)

Lahiri (1951) suggests the following method for drawing the sample with probability proportional to $\Sigma_1^n M_i$. Let T denote the total of the n largest values of M_i in the population. Draw a simple random sample of n units without replacement, and a random number r between 1 and T. If $\Sigma_1^n M_i \geq r$, retain this sample; otherwise, repeat the procedure. Since the sum of $\Sigma_1^n M_i$ over all the possible $_N C_n$ samples equals $(N-1)^C$ $(n-1)M_0$, the probability of drawing a specified sample is

$$P_k = \frac{\sum_1^n M_i}{\binom{N-1}{n-1}M_0} = \frac{\overline{m}}{\binom{N}{n}\overline{M}}, \tag{7.39}$$

where k runs from 1 to $_N C_n$.

The estimator for $\overline{Y}$ considered with this procedure is $\hat{\overline{Y}}_{SS} = \overline{M}(\overline{y}/\overline{m})$. As shown in Appendix A7, this estimator is unbiased for $\overline{Y}$ and for large n has the same variance as (7.21) for $\hat{\overline{Y}}_R$. Thus, in addition to being unbiased, $\hat{\overline{Y}}_{SS}$ has smaller variance than $\overline{y}$ provided y_i have a high positive correlation with the cluster sizes M_i. An estimator for the exact variance of $\hat{\overline{Y}}_{SS}$ was obtained by Des Raj (1968), but it may take negative values.

In an alternative procedure suggested by Midzuno (1951), the first unit of the sample is selected with probability proportional to M_i and the remaining $(n-1)$ units are drawn from the $(N-1)$ units randomly without replacement. This approach also results in selecting the sample with probability proportional to $\Sigma_1^n M_i$.

Randomization approach

For the procedure suggested by J.N.K. Rao et al. (1962), the relative size of a unit or a measure of the size is denoted by z_i. In this approach, the population is first divided randomly into n groups of sizes N_g, $g = (1,...,n)$, $N = \Sigma_1^n N_g$. Denoting the total of the gth group by Y_g, the population total becomes $Y = \Sigma_1^g Y_g$. The relative sizes of the population units are denoted by z_i, $i = (1, 2,...,N)$, and the total of the N_g relative sizes of the gth group by Z_g. Now, the probabilities $p_i = z_i/Z_g$ are assigned to the N_g units of the gth group and one unit is selected from each of the n groups with these probabilities.

Let y_g and z_g denote the observation of the sampled unit and its original probability. For a given grouping of the units, let $p_g = z_g/Z_g$. Now, $\hat{Y}_g = (y_g/p_g)$ is unbiased for the total Y_g of the gth group. Hence, $\hat{Y}_{HRC} = \Sigma_1^n \hat{Y}_g$ is unbiased for the population total Y. The variance of this estimator can be obtained from the approach in Appendix A3. The above authors and Cochran (1977, p. 267), derive this variance and its estimator. When $N_g = N/n$ is an integer, they are given by

$$V(\hat{Y}_{RHC}) = \frac{N-n}{(N-1)n} \sum_1^N z_i\left(\frac{y_i}{z_i} - Y\right)^2 \qquad (7.40)$$

and

$$v(\hat{Y}_{RHC}) = \frac{N-n}{N(n-1)} \sum_1^n Z_g\left(\frac{y_g}{z_g} - \hat{Y}_{RHC}\right)^2. \qquad (7.41)$$

As an illustration, for the four units described in Sections 7.11 and 7.12, z_i are the same as the relative sizes (0.375, 0.30, 0.25, 0.075), and $y_i = (105, 80, 45, 10)$. A random division of the units into two groups of equal sizes may result, for example, in the groups with units (1, 4) and (2, 3). The z_i for the first group are (0.375, 0.075) and their sum is $Z_1 = 0.45$. Hence, the probabilities of selection for the two units of this group are $0.375/0.45 = 0.83$ and $0.075/0.45 = 0.17$. One unit is chosen from this group with these probabilities. Similarly, the probabilities of selection for the units of the second group are $0.30/0.55 = 0.545$ and $0.25/0.55 = 0.455$, and one unit is chosen from this group with these probabilites. In general, since $z_i = M_i/M_0$ for the cluster units, the variance in (7.40) will be small if the correlation between y_i and M_i is positively large.

The estimator for $\bar{Y}$ is obtained from dividing $\hat{Y}_{RHC}$ by N. The variance of the resulting estimator and its variance estimator are obtained by dividing (7.40) and (7.41) by N^2.

Comparisons of $\hat{Y}_{RHC}$ with the ratio estimator in Section 7.8 and the PPSS estimator through a suitable model are presented in Chapter 9.

Selection with probabilities proportional to size and replacement (pps)

Following the procedure in Section 7.11 and replacing a unit that has appeared in the sample, one can select a sample of n units with probabilities proportional to the relative sizes u_i. Denoting y_i/u_i by r_i, an estimator for Y is $\hat{Y}_{pps} = \Sigma_1^n r_i/n$. Since the expectation of r_i is $\Sigma_1^N u_i(y_i/u_i) = Y$, $E(\hat{Y}_{pps}) = Y$. Thus, this estimator is unbiased for Y.

The variance of $\hat{Y}_{pps}$ and the estimator of variance are derived in Appendix A7. The variance is given by $V(\hat{Y}_{pps}) = s_r^2/n$, where $s_r^2 = \overset{N}{\underset{1}{\Sigma}} u_i (r_i - Y)^2$. Thus, the variance of this unbiased estimator also becomes small as y_i is highly correlated with u_i. The estimator of this variance is given by $v(\hat{Y}_{pps}) = s_r^2/n$, where $s_r^2 = \Sigma_1^n (r_i - \hat{Y}_{pps})^2/(n-1)$.

Although the procedure is easy to implement, selecting the units with replacement is not practical, since no additional information is obtained by including a unit more than once in the sample. If the sample units, however, are selected with probabilities u_i and replacement, Pathak (1962) shows that the mean $\hat{Y}_d = \Sigma_1^d r_i/d$ of the $d(\leq n)$ distinct sample units obtained by removing the duplications is unbiased for Y and has smaller variance than $\hat{Y}_{pps}$. As an illustration, when two distinct units appear in a sample of size three selected with replacement, that is, $n = 3$ and $d = 2$, he shows that $r_1 = y_1/u_1$, $r_2 = y_2/u_2$, and $r_{12} = (y_1 + y_2)/(u_1 + u_2)$ are unbiased for Y and have smaller variances than $\hat{Y}_{pps}$.

Exercises

7.1. For the 54 establishments in Table 7.1, $S^2 = 15.94$. To estimate the average number of establishments per county, compare the precisions of a sample of three clusters and a simple random sample of 18 counties.

7.2. If clusters 2, 3, and 8 are drawn into the sample from the nine clusters of Table 7.1, (a) estimate the average employment per county and (b) find its standard error. Use the summary figures in Table 7.2.

7.3. With the sample clusters in Example 7.3, estimate the precision of cluster sampling relative to simple random sampling for the estimation of the averages of the (a) employment and (b) number of establishments.

7.4. For estimating the proportion of the 54 counties with more than 10,000 employees, with the data in Table 7.4 and a sample of three clusters, compare the variances of (a) the sample proportion in (7.26) and (b) the ratio type estimator in (7.27).

7.5. (a) From the data in Table 7.4, find the proportion and total number of the 54 counties that have more than 10,000 employees. (b) If clusters 2, 3, and 8 are drawn into the sample from the nine clusters, estimate the above proportion through the sample mean of the proportions and the ratio method, and (c) compare the sample standard errors of the two estimates.

7.6. If clusters 2, 3, and 8 are selected in a sample of three clusters, as in Exercise 7.5, (a) estimate by both approaches the total number of counties that have more than 10,000 employees and (b) find the standard errors of the estimates.

7.7. *Project.* Select all the samples of size three randomly without replacement from the nine clusters of unequal sizes in Table 7.3, and compute the ratio estimate in (7.17) from each sample. (a) From these estimates, find the expectation, bias, and MSE of the ratio estimator, and compare this exact MSE with the approximate expression in (7.18). (b) Compute the variance estimate in (7.19) from each sample, and find the bias of this estimator for the exact variance or MSE of the ratio estimator in (7.17).

7.8. For samples of sizes $n_1 = 2$, $n_2 = 1$, and $n_3 = 2$ selected randomly from the three strata of Section 7.10, find the variance of the estimators for the population total and mean per unit.

7.9. For the stratum with the four clusters 5 through 8 in Table 7.3, the sizes, number of establishments, and employment figures are $M_i = (4, 5, 5, 7)$, $x_i = (6.5, 6.4, 9.1, 19.5)$, and $y_i = (70.8, 70.6, 118.5, 273.2)$. Consider the probabilities $u_i = (0.2, 0.25, 0.25, 0.3)$ for selecting these units into the sample. For selecting a sample of two units from this stratum as described in Section 7.11, find the probabilities in (7.31) and (7.32).

7.10.   Find the variance of the Horvitz–Thomson estimator of $\bar{\bar{Y}}$ for the procedure in Exercise 7.9 and compare it with the estimator obtained from a simple random sample of two units.

7.11.   If the second and third units are selected into the sample through the procedure in Exercise 7.9, estimate the S.E. of the Horvitz–Thomson estimators for Y and $\bar{\bar{Y}}$ through both (7.37) and (7.38).

7.12.   Following the approach of Des Raj (1968), express $\bar{\bar{Y}}^2$ as $(\Sigma_i^N y_i^2 + \Sigma_{i \neq j}^N \Sigma^N y_i y_j)/N^2$ and find an expression for the estimator of $V(\hat{\bar{Y}}_{SS})$.

7.13.   Show that $V(\hat{Y}_{pps})$ and $v(\hat{Y}_{pps})$ can be expressed as $\Sigma_i^N \Sigma_{i<j}^N u_i u_j (r_i - r_j)^2/n$ and $\Sigma_i^n \Sigma_{i<j}^n (r_i - r_j)^2/n^2(n-1)$, respectively.

7.14.   Find an estimator for $V(\hat{Y}_{pps})$ directly from its expression in Section 7.13.

Appendix A7

Alternative expression for S^2

The numerator of (7.5) can be expressed as

$$
\sum_1^N \sum_1^M (y_{ij} - \bar{\bar{Y}})^2 = \sum_1^N \sum_1^M [(y_{ij} - \bar{y}_i) + (\bar{y}_i - \bar{\bar{Y}})]^2
$$

$$
= \sum_1^N \sum_1^M (y_{ij} - \bar{y}_i)^2 + M \sum_1^N (\bar{y}_i - \bar{\bar{Y}})^2
$$

$$
= (M - 1)\sum_1^N s_i^2 + M\sum_1^N (\bar{y}_i - \bar{\bar{Y}})^2,
$$

where s_i^2 is the variance within the ith cluster. Note that the cross-product term in the square brackets of the first expression vanishes. In the *Analysis of Variance* terminology, the left-hand side is the *total SS* (sum of squares). The two expressions on the right-hand side are the *within SS* and *between SS*, respectively.

Now, from (7.5),

$$S^2 = \frac{N(M-1)}{NM-1}S^2_{wsy} + \frac{N-1}{NM-1}S^2_b.$$

Estimator for S^2

The variance among the nM sample observations can be expressed as

$$s^2 = \frac{\sum_1^n \sum_1^M (y_{ij} - \bar{\bar{y}})^2}{nM-1}$$

$$= \frac{\sum_1^n \sum_1^M (y_{ij} - \bar{y}_i)^2 + M\sum_1^n (\bar{y}_i - \bar{\bar{y}})^2}{nM-1}$$

$$= \frac{n(M-1)}{nM-1}s^2_w + \frac{n-1}{nM-1}s^2_b,$$

where $s^2_w = (\sum_1^n s^2_i)/n$ is the sample *within* MS with $n(M-1)$ d.f.

The sample *between* MS and *within* MS are unbiased for S^2_b and S^2_w, respectively. However, s^2 is not unbiased for the population variance S^2 in (7.5). From (7.6), its unbiased estimator is given by

$$\hat{S}^2 = \frac{N(M-1)}{NM-1}s^2_w + \frac{N-1}{NM-1}s^2_b,$$

which becomes $[(M-1)s^2_w + s^2_b]/M$ for large N.

The relative precision in (7.13) can be estimated by substituting $\hat{S}^2$ and s^2_b for S^2 and S^2_b, respectively.

Bias of the ratio estimator for the mean

The bias of $\hat{\bar{\bar{Y}}}_R = \bar{y}/\bar{m}$ is

$$B(\hat{\bar{\bar{Y}}}_R) = E(\hat{\bar{\bar{Y}}}_R) - \bar{\bar{Y}} = E[(\bar{y} - \bar{\bar{Y}}\bar{m})/\bar{m}]$$

$$= E[(\bar{y} - \bar{\bar{Y}}\bar{m})/\bar{M}(1+\delta)],$$

where $\delta = (\bar{m} - \bar{M})/\bar{M}$. By expanding $(1 + \delta)^{-1}$ as $1 - \delta + \delta^2 - \delta^3 + \ldots$, and ignoring terms of order $1/n^2$ and smaller, the bias would be of order $1/n$. For large n, ignoring terms of order $1/n$ also, the bias is $E[(\bar{y} - \bar{\bar{Y}}\bar{m})/\bar{M}] = 0$.

Variance of the ratio estimator

From the large sample approximation in Appendix A7, the MSE of $\hat{\bar{Y}}_R$ becomes the same as its variance

$$V(\hat{\bar{Y}}_R) = E(\bar{y} - \bar{\bar{Y}}\bar{m})^2/\bar{M}^2.$$

The term inside the parenthesis is the mean $\bar{d}$ of $d_i = y_i - \bar{\bar{Y}}m_i$. The population mean of d_i and hence of $\bar{d}$ is equal to zero. The variance $\hat{\bar{Y}}_R$ can be expressed as $V(\hat{\bar{Y}}_R) = V(\bar{d})/\bar{M}^2 = [(1 - f)/n]S_d^2$, where

$$S_d^2 = \frac{1}{N-1}\sum_1^N(y_i - \bar{\bar{Y}}m_i) = \frac{1}{N-1}\sum_1^N[(y_i - \bar{Y}) - \bar{\bar{Y}}(m_i - \bar{M})]^2$$

$$= S_y^2 + \bar{\bar{Y}}^2S_m^2 - 2\bar{\bar{Y}}S_{my} = S_y^2 + \bar{\bar{Y}}^2S_m^2 - 2\bar{\bar{Y}}\rho S_m S_y.$$

From the sample,

$$s_d^2 = \frac{1}{n-1}\sum_1^n(y_i - \hat{\bar{\bar{Y}}}_R m_i)^2$$

$$= s_y^2 + \hat{\bar{\bar{Y}}}_R^2 s_m^2 - 2\hat{\bar{\bar{Y}}}_R s_{my} = s_y^2 + \hat{\bar{\bar{Y}}}_R^2 s_m^2 - 2\hat{\bar{\bar{Y}}}_R r s_m s_y,$$

where $r = s_{my}/s_m s_y$.

Hence, $V(\hat{\bar{Y}}_R)$ may be approximately estimated from

$$v(\bar{\bar{Y}}_R) = \frac{(1-f)}{n\bar{m}^2}s_d^2.$$

Expectation and variance of the Horvitz–Thompson estimator

This estimator $\hat{Y}_{HT}$ in (7.34) can be expressed as $\Sigma_1^N\delta_i(y_i/\phi_i)$. The random variable δ_i takes the value one when a unit is selected into the sample and zero otherwise. When the ith population unit is selected into the sample with probability ϕ_i, $P(\delta_i = 1) = \phi_i$ and $P(\delta_i = 0) = 1 - \phi_i$. As a result, $E(\delta_i) = 1(\phi_i) + 0(1 - \phi_i) = \phi_i$. Thus, $E(\hat{Y}_{HT}) = \Sigma_1^N E(\delta_i)(y_i/\phi_i) = \Sigma_1^N \phi_i(y_i/\phi_i) = Y$.

The variance of $\hat{Y}_{HT}$ can be expressed as $V(\hat{Y}_{HT}) = E(\hat{Y}_{HT} - Y)^2 = E(\hat{Y}_{HT})^2 - 2YE(\hat{Y}_{HT}) + Y^2 = E(\hat{Y}_{HT})^2 - Y^2$. Now,

$$\hat{Y}_{HT}^2 = \sum_1^n \left(\frac{y_i}{\phi_i}\right)^2 + \sum_{i \neq j}^n \sum^n \frac{y_i}{\phi_i} \frac{y_j}{\phi_j} = \sum_1^N \delta_i \left(\frac{y_i}{\phi_i}\right)^2 + \sum_{i \neq j}^N \sum^N \delta_{ij} \frac{y_i}{\phi_i} \frac{y_j}{\phi_j}.$$

The random variable δ_{ij} takes the value one when the ith and jth units are selected into the sample and zero otherwise, with $P(\delta_{ij} = 1) = \phi_{ij}$ and $P(\delta_{ij} = 0) = 1 - \phi_{ij}$. Thus,

$$E(\hat{Y}_{HT}^2) = \sum_1^N \phi_i \left(\frac{y_i}{\phi_i}\right)^2 + \sum_{i \neq j}^N \sum^N \phi_{ij} \frac{y_i}{\phi_i} \frac{y_j}{\phi_j}.$$

Now, the variance of $\hat{Y}_{HT}$ is obtained by subtracting $Y^2 = \Sigma_i y_i^2 + \Sigma_{i \neq j} \Sigma y_i y_j$ from this expression.

Expectation and variance of the PPSS estimator

The expectation of $\hat{\bar{Y}}_{SS}$ is $\bar{M} \Sigma P_k(\bar{y}/\bar{m})_k = \Sigma \bar{y}_k /_N C_n = \bar{Y}$. Thus, $\hat{\bar{Y}}_{SS}$ is unbiased for $\bar{Y}$.

Further, the expectation of $\hat{\bar{Y}}_{SS}^2$ is $\bar{M}^2 \Sigma P_k(\bar{y}/\bar{m})^2 = \bar{M} \Sigma(\bar{y}^2/\bar{m})/_N C_n = \bar{M} E(\bar{y}^2/\bar{m})$, where E now stands for expectation with respect to simple random sampling without replacement. Thus, $V(\hat{\bar{Y}}_{SS}) = \bar{M} E(\bar{y}^2/\bar{m}) - \bar{Y}^2$. The large sample approach for the ratio estimator leads to

$$V(\hat{\bar{Y}}_{SS}) = \frac{(1 - f)}{n}(S_y^2 + R^2 S_m^2 - 2R\rho S_y S_m),$$

which is the same as (7.21).

Variance of the pps estimator and the variance estimator

Since the sample is selected with replacement, $V(\hat{Y}_{pps})$ is the same as S_r^2/n, where $S_r^2 = V(r_i)$. Since $E(r_i) = Y$, $V(r_i) = E(r_i^2) - Y^2$. Now, $E(r_i^2) =$

$\Sigma_1^N u_i r_i^2 = \Sigma_1^N(y_i^2/u_i)$. Therefore, $V(r_i) = \Sigma_1^N(y_i^2/u_i) - Y^2 = \Sigma_1^N(y_i - u_i Y)^2/u_i = \Sigma_1^N u_i(r_i - Y)^2$.

To find the estimators for $V(r_i)$ and $V(\hat{Y}_{pps})$, note that $\Sigma_1^n(r_i - \hat{Y}_{pps})^2 = \Sigma_1^n[(r_i - Y) - (\hat{Y}_{pps} - Y)]^2 = \Sigma_1^n(r_i - Y)^2 - n(\hat{Y}_{pps} - Y)^2$. Thus, $E\Sigma_1^n(r_i - \hat{Y}_{pps})^2 = n\Sigma_1^N u_i(r_i - Y)^2 - nV(\hat{Y}_{pps}) = n(n - 1)V(\hat{Y}_{pps})$. Hence, $s_r^2 = \Sigma_1^n(r_i - \hat{Y}_{pps})^2/(n - 1)$ is unbiased for S_r^2 and $v(\hat{Y}_{pps}) = s_r^2/n$ is unbiased for $V(\hat{Y}_{pps})$.

CHAPTER 8

Sampling in Two Stages

8.1   Introduction

For the single-stage cluster sampling discussed in the last chapter, observations are obtained from all the elements of each selected cluster. When the clusters contain a large number of elements, estimates for the population means, totals, and proportions are usually obtained from subsamples of elements in the clusters. Mahalanobis (1946) describes this procedure as the **two-stage sampling**. For this approach, clusters are the first-stage units, which are also known as **primary sampling units** (PSUs). The elements are the subunits or **secondary sampling units** (SSUs).

For an agricultural survey, districts or enumeration areas can be the PSUs, and the villages can be considered to be the SSUs. For estimation of employment rates and related characteristics in a geographical region, the counties and city blocks in the region are usually considered to be the PSUs and SSUs, respectively. For health-related surveys, counties or a group of counties can be considered as the PSUs and the households or hospitals as SSUs. For large-scale surveys, stratification may precede selection of the sample at any stage.

This chapter presents some of the frequently used procedures for selecting the first- and second-stage units with equal or unequal probabilities. For some surveys, more than two stages of selection is considered.

8.2   Equal size first-stage units

The notation for this case is similar to that in Sections 7.2 through 7.6 except that the cluster totals are denoted by Y_i. Let y_{ij}, $i = 1, ..., N$ and $j = 1, ..., M$, denote the observations of the M second-stage units on each of the N first-stage units. The totals of the ith first-stage unit and the population are $Y_i = \Sigma^M y_{ij}$ and $Y = \Sigma^N Y_i$. The mean of the ith first-stage unit is $\bar{Y}_i = Y_i/M$. The population mean per primary unit and per subunit or element are $\bar{Y} = Y/N$ and $\bar{\bar{Y}} = Y/M_o = (\Sigma^N Y_i)/N$, where $M_o = NM$. The variances among the means of the first-stage units and within the elements of the second-stage units, respectively, are

$$S_1^2 = \sum_1^N (\bar{Y}_i - \bar{\bar{Y}})^2/(N-1) \quad \text{and} \quad S_2^2 = \sum_1^N S_{2i}^2/N, \quad (8.1)$$

where $S_{2i}^2 = \Sigma_1^M (\bar{y}_{ij} - \bar{Y}_i)^2/(M-1)$. Note that S_{2i}^2 is the variance within the ith primary unit with $(M-1)$ d.f. and S_2^2 is the pooled within variance with $N(M-1)$ d.f.

At the first stage, a sample of n primary units is selected, and at the second stage a sample of m secondary units is selected from each of the selected primary units. The mean of the m subsampled units of the ith primary unit and the overall mean of the nm subsampled units, respectively, are

$$\bar{y}_i = \sum_1^m y_{ij}/m \quad \text{and} \quad \bar{\bar{y}} = \sum_1^n \sum_1^m y_{ij}/nm = \sum^n \bar{y}_i/n. \quad (8.2)$$

The sample variances among the primary units and the secondary units, respectively, are

$$s_1^2 = \sum_1^n (\bar{y}_i - \bar{\bar{y}})^2/(n-1) \quad \text{and} \quad s_2^2 = \sum_1^n s_{2i}^2/n, \quad (8.3)$$

where $s_{2i}^2 = \Sigma_1^m (y_{ij} - \bar{y}_i)^2/(m-1)$ is the variance within the ith primary unit.

8.3   Estimating the mean

Estimator for the mean and its variance

For each of the primary units selected at the first stage, $\bar{y}_i$ is unbiased for the mean $\bar{Y}_i$. As shown in Appendix A8, $\bar{\bar{y}}$ is unbiased for $\bar{\bar{Y}}$ and its variance is given by

$$V(\bar{y}) = (1 - f_1)S_1^2/n + (1 - f_2)S_2^2/nm, \tag{8.4}$$

where $f_1 = n/N$ and $f_2 = m/M$ are the sampling fractions at the first and second stages.

Since only m of the M secondary units are sampled, the variance increased from single-stage sampling by the amount in the second term. In practice, the optimum values of n and m for minimizing the variance or cost of sampling are found.

Example 8.1. Establishments per county: For the number of establishments in Table 7.1 for the nine clusters with six counties in each, $\bar{\bar{Y}} = 3.02$, $S_1^2 = 2.63$, and $S_2^2 = 15.96$. If $n = 3$ and $m = 3$, from (8.4), $V(\bar{y}) = (6/27)(2.63) + 15.96/18 = 1.47$.

Since the population variance S^2 is equal to 15.94, for the alternative procedure of selecting a simple random sample of size $nm = 9$ from the 54 counties, the variance is $V(\bar{y}_{srs}) = (5/54)(15.94) = 1.48$. In this illustration, there is very little difference in the precision of $\bar{y}$ and $\bar{y}_{srs}$. However, there can be saving in costs for cluster sampling.

Estimation of the variance

The sample variance s_{2i}^2 is unbiased for S_{2i}^2, and hence s_2^2 is unbiased for S_2^2. This result can be used for estimating the second term of (8.4). As shown in Appendix A8, $s_1^2 - (1-f_2)s_2^2/m$ is unbiased for S_1^2. Using this estimator for S_1^2 and s_2^2 for S_2^2, an unbiased estimator for the variance in (8.4) is

$$v(\bar{y}) = (1 - f_1)s_1^2/n + f_1(1 - f_2)s_2^2/nm. \tag{8.5}$$

Example 8.2. Standard error for estimating employment per county: In a random sample of size three from the nine clusters described in Table 7.1, clusters 3, 5, and 9 appeared. In random samples of size three from each of these clusters, counties (2, 4, 5), (1, 4, 6), and (2, 3, 6), respectively, were drawn. The numbers of establishments and the persons employed in them are presented in Table 8.1.

Table 8.1. A sample of three clusters of equal sizes ($M = 6$) and samples of three counties from the selected clusters.

Clusters (PSUs)	Counties (SSUs)	Establishments	Employments
	2	1.9	341.2
3	4	11.2	14.4
	5	5.1	7.8
	1	2.0	20.7
5	4	0.5	3.8
	6	1.6	17.8
	2	5.5	90.0
9	3	6.2	69.9
	6	13.1	170.5

For the establishments, the sample means $\bar{y}_i$ are 6.07, 1.37, and 8.27. The sample variances s_{2i}^2 are 22.32, 0.60, and 17.64. From these figures, $\bar{y} = (6.07 + 1.37 + 8.27)/3 = 5.24$. Note that the actual mean is $\bar{Y} = 3.02$, as seen in Example 8.1. Further, $s_1^2 = 2.42$ and $s_2^2 = (22.32 + 0.60 + 17.64)/3 = 13.52$. Now, from (8.5), $v(\bar{y}) = (2/9)(2.42) + 13.52/54 = 0.79$. The actual variance of $\bar{y}$ as seen in Example 8.1 is 1.47.

8.4   Sample size determination

Continuing with Example 8.1, if $n = 4$ and $m = 2$, from (8.4), $V(\bar{y}) = 5(2.63)/36 + 15.96/12 = 1.03$. On the other hand, if $n = 2$ and $m = 4$, $V(\bar{y}) = 7(2.63)/18 + 15.96/24 = 1.6878$. In both cases, $V(\bar{y}_{\text{srs}}) = 46(15.94)/432 = 1.6973$. These illustrations show that proper choice of n and m can result in a high precision for cluster sampling.

The cost of sampling the nm elements may be of the form $E = enm$, where e is the expense for collecting information from a selected unit. The sample sizes can be determined for a specified cost and the variance in (8.4).

As an illustration, if $e = \$100$ and $E = \$1200$, $nm = 12$. Now, if $S_1^2 = 2.63$ and $S_2^2 = 15.96$ as in Example 8.1, from (8.4), the variance of $\bar{y}$ equals $2.63/n - 2.63/9 + 15.96/12 - 15.96/(6n) = 1.0378 - 0.03/n$. Thus, if $(n, m) = (4, 3)$ or $(3, 4)$, the variance will be close to 1.03. Other types of cost functions suitable to the application can be considered to examine the required sample sizes.

8.5 Unequal size primary units

The notation for this case is the same as in Section 7.7, except that the total of the ith cluster is denoted by Y_i. When the ith primary unit contains M_i secondary units, its total and mean are $Y_i = \Sigma^{M_i} y_{ij}$ and $\bar{Y}_i = Y_i/M_i$. The population mean per primary and secondary unit, respectively, are $\bar{Y} = \Sigma^N \bar{Y}_i/N$ and $\bar{\bar{Y}} = \Sigma^N Y_i/M_0 = \Sigma^N M_i \bar{Y}_i/M_0$, where $M_0 = \Sigma^N M_i$ is the total number of secondary units. The variance within the ith primary unit is $S_{2i}^2 = \Sigma^{M_i}(y_{ij} - \bar{Y}_i)^2/(M_i - 1)$.

As before, at the first stage, a sample of n primary units is selected without replacement. At the second stage, a sample of size m_i is selected randomly without replacement from the M_i secondary units of the selected primary unit. The mean of this sample is $\bar{y}_i = \Sigma^{m_i} y_{ij}/m_i$, which is unbiased for $\bar{Y}_i$. Its variance $s_{2i}^2 = \Sigma^{m_i}(y_{ij} - \bar{y}_i)^2/(m_i - 1)$ is unbiased for S_{2i}^2.

Estimators for the population total and mean

An unbiased estimator of Y_i is $\hat{Y}_i = M_i \bar{y}_i$. As shown in Appendix A8, $\hat{Y} = (N/n)\Sigma^n \hat{Y}_i$ is unbiased for Y and its variance is given by

$$V(\hat{Y}) = [N^2(1 - f_1)/n]\sum^N (Y_i - \bar{Y})^2/(N - 1)$$

$$+ (N/n)\sum^N M_i^2(1 - f_{2i})S_{2i}^2/m_i. \qquad (8.6)$$

An unbiased estimator of $\bar{Y}$ is $\hat{\bar{Y}} = \hat{Y}/N = (\Sigma^n \hat{Y}_i)/n$. Its variance is obtained by dividing (8.6) by N^2. Similarly, $\hat{Y}/M_0$ is unbiased for $\bar{\bar{Y}}$ and its variance is obtained by dividing (8.6) by M_0^2. If the totals Y_i vary with the sizes, the first term of (8.6) becomes large. The ratio method to adjust the estimator for the sizes is described in Section 8.6.

Example 8.3. Establishments: Consider a random sample of size three from the clusters described in Table 7.3. For second-stage sampling, consider $m_i = 3$ counties for the first two and last two clusters selected at the first stage and $m_i = 2$ counties for the remaining five clusters selected at the first stage.

For the establishments, the first term of (8.6) equals $9(6)(16.42)^2/3 = 4853.1$ and the second term equals 5065.1. Combining these figures, $V(\hat{Y}) = 9918.1$ and $V\hat{\bar{Y}} = 9918.1/(54)^2 = 3.4$.

Estimation of the variance

From the second and third expressions in Appendix A8, an unbiased estimator of (8.6) is given by

$$v(\hat{Y}) = N^2[(1 - f_1)/n]\sum_{}^{n}(\hat{Y}_i - \hat{\bar{Y}})^2/(n - 1)$$

$$-(N/n)^2(1 - f_1)\left[\sum_{}^{n}M_i^2(1 - f_{2i})s_{2i}^2/m_i\right]$$

$$+ (N/n)^2\sum_{}^{n}M_i^2(1 - f_{2i})s_{2i}^2/m_i$$

$$= N^2[(1 - f_1)/n]\sum_{}^{n}(\hat{Y}_i - \hat{\bar{Y}})^2/(n - 1)$$

$$+ (N/n)\sum_{}^{n}M_i^2(1 - f_{2i})s_{2i}^2/m_i. \qquad (8.7)$$

To estimate the variance of $\hat{\bar{\bar{Y}}} = \hat{Y}/M_0$, divide (8.7) by M_0^2.

Example 8.4. Standard error from the sample: In a random sample of size three from the clusters of unequal size described in Table 7.3, clusters 1, 4, and 8 were drawn. In random subsamples of sizes 3, 2, and 3 from the selected clusters, counties (3, 4, 10), (2, 4), and (2, 4, 5), respectively, were drawn. Figures for the establishments and employments for these samples are presented in Table 8.2. For the establishments, the subsample means $\bar{y}_i$ and variances s_i^2 for the sample clusters are (7.77, 2.75, 4.57) and (137.4, 6.125, 8.9), respectively. Estimates $\hat{Y}_i$ for the three selected clusters are $10(7.77) = 77.7$, $4(2.75) = 11$, and $7(4.57) = 31.99$. Hence, $\hat{Y} = (9/3)(77.7 + 11 + 31.99) = 362.1$, $\hat{\bar{Y}} = 362.1/9 = 40.23$, and $\hat{\bar{\bar{Y}}} = 362.1/54 = 6.7$. Note that $Y = 162.9$ and $\bar{\bar{Y}} = 162.9/54 = 3.02$.

The above estimates for the total and mean are about twice their actual values. This has happened because the two largest clusters, 1 and 8, appeared in the sample. Estimators with smaller variances can be obtained if the PSUs are stratified according to their sizes. If supplementary information is available, the ratio and regression procedures presented in Chapters 9 and 10 can be beneficial.

Since $\Sigma_1^3(\hat{Y}_i - \hat{\bar{Y}})^2/2 = 1163.1$, the first term of (8.7) equals 20,936.6. With the second-stage variances, the second term equals 9940.7. Thus, $v(\hat{Y}) = 30,877.3$ and S.E.$(\hat{Y}) = 175.7$. For the estimation of $\bar{\bar{Y}}$, S.E.$(\bar{\bar{Y}}) = 175.7/54 = 3.25$.

Table 8.2. Sample of clusters of unequal size and samples of counties.

Clusters (PSUs)	Cluster Sizes M_i	Counties (SSUs)	Establishments	Employments
1	10	3	21.3	341.2
		4	1.2	14.4
		10	0.8	10.1
4	4	2	1.0	13.7
		4	4.5	84.7
8	7	2	3.1	52.5
		4	8.0	132.0
		5	2.6	34.2

Hansen and Hurwitz (1949) and Cochran (1977, p. 313) determine the sample sizes for the two stages with a suitable cost function.

8.6 Ratio adjustments for the sizes

The first term of the variance in (8.6) becomes large if Y_i vary much from each other. In many applications Y_i depends on the size M_i. In such cases, the ratio method can be used for estimating Y with a small MSE. The ratio-type estimator for Y is

$$\hat{Y}_R = M_0 \left(\sum_{}^{n} \hat{Y}_i \right) \bigg/ \left(\sum_{}^{n} M_i \right) = N\overline{M} \left(\sum_{1}^{n} \hat{Y}_i / n \right) \bigg/ \left(\sum_{}^{n} M_i / n \right)$$

$$= N\overline{M} \left(\sum_{}^{n} M_i \bar{y}_i / n \right) \bigg/ \overline{m},$$

(8.8)

where $\overline{M} = M_0/N$ is the average size of the primary units and $\overline{m} = \Sigma^n M_i / n$ is the mean of the sizes of the cluster selected at the first stage. The estimator of the population mean is given by $\hat{\overline{Y}}_R = \hat{Y}_R / M_0$. For large n, the bias of $\hat{Y}_R$ in (8.8) becomes small, and from Appendix A8 its variance is approximately given by

$$V(\hat{Y}_R) = N^2 [(1 - f_1)/n] \sum^{N} M_i^2 (\overline{Y}_i - \overline{\overline{Y}})^2 / (N - 1)$$

$$+ (N/n) \left[\sum^{N} M_i^2 (1 - f_{2i}) S_{2i}^2 / m_i \right].$$

(8.9)

The first term of this expression will be small if Y_i is close to $\bar{\bar{Y}} M_i$, that is, if Y_i is correlated with M_i. Notice that the second terms of (8.6) and (8.9) are the same.

Starting with the expectation of $\Sigma^n M_i^2 (\bar{y}_i - \hat{\bar{Y}}_R)^2/(n-1)$ and adopting the procedure in Appendix A8, an estimator for $V(\hat{Y}_R)$ is given by

$$v(\hat{Y}_R) = N^2[(1 - f_1)/n] \sum^n M_i^2 (\bar{y}_i - \hat{\bar{Y}}_R)^2/(n-1)$$

$$+ (N/n) \sum^n M_i^2 (1 - f_{2i}) s_{2i}^2/m_i. \qquad (8.10)$$

Note that the second terms in (8.7) and (8.10) are the same.

Example 8.5. Ratio estimation for the establishments: With the samples drawn in Example 8.4, for the number of establishments, $\hat{Y}_R = 54$ $(120.69)/21 = 310.35$ and $\hat{\bar{Y}}_R = 310.35/54 = 5.75$. Since $\Sigma^n M_i^2 (\bar{y}_i - \hat{\bar{Y}}_R)^2 = 592.5$, the first term of (8.10) equals 5332.3. The second term as in Example 8.4 equals 9940.7. Thus, $v(\hat{Y}_R) = 15,273$ and S.E.$(\hat{Y}_R) = 123.6$. Hence, S.E.$(\hat{\bar{Y}}_R) = 2.29$.

The above ratio estimators for the total and mean differ only slightly from $\hat{Y}$ and $\bar{\bar{Y}}$ found in Example 8.4. For the estimation of these quantities in this illustration, the ratio method has helped only by a small amount.

8.7   Proportions and totals

Let C_i and P_i denote the total number and proportion of the M_i secondary units having the characteristic of interest. The total number and proportion for the population are $C = \Sigma^N C_i$ and $P = C/M_0$. With the one-zero notation for y_{ij}, the totals and means can be expressed as $Y_i = C_i$, $\bar{Y}_i = P_i$, and $\bar{\bar{Y}} = P$. Similarly, let c_i and $p_i = c_i/m_i$ denote the total number and proportion of the m_i secondary units of the sample having the attribute.

Equal size PSUs

When $M_i = M$ and $m_i = m$, from (8.2), an unbiased estimator for P is

$$p = \sum_1^n p_i/n = \sum^n c_i/nm. \qquad (8.11)$$

The variance of this estimator can be obtained from (8.4). From (8.5), the estimator of variance is

$$v(p) = \frac{(1-f_1)}{n} \frac{\sum^n (p_i - p)^2}{n-1} + \frac{f_1(1-f_2)}{n^2(m-1)} \sum_1^n p_i(1-p_i). \quad (8.12)$$

An unbiased estimator of C is given by $M_0 p$. For the sample variance of this estimator, multiply (8.12) by M_0^2.

Example 8.6. Employment in the counties: Observations from a sample of $n = 3$ clusters and subsamples of $m = 3$ counties from the selected clusters are presented in Table 8.1. Numbers of counties with employment of at least 20,000 persons are $c_1 = 1$, $c_2 = 1$, and $c_3 = 3$. Thus, $p_1 = 1/3$, $p_2 = 1/3$, and $p_3 = 1$. Estimates for the proportion and total number of counties with this characteristic are $p = (5/9) = 0.56$ and $\hat{C} = 54(5/9) = 30$. From Table 7.1, the actual total and proportion are $C = 25$ and $P = 25/54 = 0.46$.

The first and second terms of (8.12) equal $(16/729)$ and $3/729$. Hence $v(p) = 19/729 = 0.0261$, S.E.$(p) = 0.16$, and S.E. $(\hat{C}) = 54(0.16) = 9$.

Unequal size PSUs

When M_i are unequal, from Section 8.5, an unbiased estimator for the total is

$$\hat{C} = \frac{N}{n} \sum_1^n M_i p_i. \quad (8.13)$$

Its variance can be obtained from (8.6). This variance will be small if C_i do not differ much from each other. From (8.7), an unbiased estimator of the variance is

$$v(\hat{C}) = \frac{N^2(1-f_1)}{n} \frac{\sum^n \left[M_i p_i - \sum^n M_i p_i/n \right]^2}{n-1}$$

$$+ \frac{N}{n} \sum_1^n M_i^2 \frac{(1-f_{2i})}{m_i - 1} p_i(1-p_i). \quad (8.14)$$

An alternative estimator

Adjusting for the sizes through the ratio method, from (8.8), an alternative estimator is given by

$$\hat{C}_R = M_0 \frac{\sum^n M_i p_i}{\sum^n M_i} = M_0 \bar{p}, \tag{8.15}$$

where $\bar{p}$ is the weighted average of the p_i. Its variance, which can be obtained from (8.9), is small if P_i are close to each other. From (8.10), an estimator for this variance is approximately given by

$$v(\hat{C}_R) = \frac{N^2(1 - f_1)}{n} \frac{\sum^n M_i^2 (p_i - \bar{p})^2}{n - 1} + \frac{N}{n} \sum_1^n M_i^2 \frac{(1 - f_{2i})}{m_i - 1} p_i (1 - p_i). \tag{8.16}$$

To estimate P when M_i are unequal, divide (8.13) or (8.15) by M_0. The corresponding estimates of variances are obtained by dividing (8.14) and (8.16) by M_0^2.

Example 8.7. S.E. for estimating employment: Sample observations from the clusters of unequal size in Table 7.3 and subsamples from the selected clusters are presented in Table 8.2. The observed numbers of counties with at least 20,000 employed persons are $c_1 = 1$, $c_2 = 1$, $c_3 = 3$. The sample proportions are $p_1 = 1/3$, $p_2 = 1/2$, $p_3 = 1$.
  Since $(\Sigma_1^3 M_i p_i)/3 = 74/18$, from (8.13), $\hat{C} = 9(74/18) = 37$. The first and second terms of (8.14) equal 120.7 and 29.3. Hence $v(\hat{C}) = 150$ and S.E.$(\hat{C}) = 12.25 = 12$. The estimate of the proportion for the above attribute is $37/54 = 0.69$. The values of 37 and 0.69 for this estimation procedure are rather large relative to the actual values 25 and 0.46.
  From (8.15), with the ratio adjustment for the sizes, $\hat{C}_R = (54/21)$ $(74/6) = 32$. The first term of (8.16) equals 134.3 and as seen above, the second term equals 29.3. Thus, $V(\hat{C}_R) = 163.6$ and S.E.$(\hat{C}_R) = 13$. The corresponding estimate of the proportion of the counties with the above attribute is $32/54 = 0.59$, which has an S.E. of 0.24. The estimates 32 and 0.59 for the total and proportion are closer to the actual values than the above estimates without the ratio adjustment for the sizes.

8.8   Unequal probability selection

When the sizes of the primary units are not the same, they can be selected with unequal probabilities through the procedures of the type described in Sections 7.11 through 7.13. From each of the selected first-stage units, samples can be selected at the second stage with equal or unequal probabilities. Selection of the first-stage units with unequal probabilities and without replacement, followed by the Horvitz-Thompson estimator, and selection of the first-stage units with unequal probabilities and with replacement are described in this section. Des Raj (1968, Chapter 6) and Cochran (1977, Chapter 11), for instance, describe additional procedures for selecting the first-stage units.

Horvitz-Thompson estimator

As described in Section 7.11, when the primary units are selected with unequal probabilities and **without** replacement, denote the inclusion probabilities for a unit and a pair of units by ϕ_i and ϕ_{ij}. The second-stage units can be selected with equal or unequal probabilities and without or with replacement. Let $\hat{Y}_i$ denote an unbiased estimator for the total Y_i of the selected primary unit obtained from the sample at the second stage, and denote its variance $V_2(\hat{Y}_i)$ by V_{2i}.

The Horvitz-Thompson estimator for the total Y now is

$$\hat{Y}_{HT} = \sum_1^n \frac{\hat{Y}_i}{\phi_i}. \tag{8.17}$$

As shown in Appendix A8, this estimator is unbiased for Y and its variance is given by

$$V(\hat{Y}_{HT}) = \sum_i^N \sum_{i<j}^N (\phi_i\phi_j - \phi_{ij})\left(\frac{Y_i}{\phi_i} - \frac{Y_j}{\phi_j}\right)^2 + \sum_1^N \frac{V_{2i}}{\phi_i}. \tag{8.18}$$

For the procedure in Section 8.2, where the samples at both the stages are drawn randomly without replacement, $\phi_i = \phi_j = n/N$, $\phi_{ij} = n(n-1)/N(N-1)$ and $V_{2i} = M^2(1-f_2)S_{2i}^2/m$. The inclusion probabilities for the procedure in Section 8.5 for the unequal size clusters also have these values, but V_{2i} in this case is equal to $M_i^2(1-f_{2i})S_{2i}^2/m_i$.

As shown in Appendix A8, an unbiased estimator for $V(\hat{Y}_{HT})$ in (8.18) is

$$v(\hat{Y}_{HT}) = \sum_i^n \sum_{i<j}^n \left(\frac{(\phi_i\phi_j - \phi_{ij})}{\phi_{ij}} \right) \left(\frac{\hat{Y}_i}{\phi_i} - \frac{\hat{Y}_j}{\phi_j} \right)^2 + \sum_1^n \frac{\hat{V}_{2i}}{\phi_i}. \qquad (8.19)$$

Selection with unequal probabilities and replacement

As described in Section 7.13 for the *pps* method, consider selecting the n first-stage units with probabilities u_i and replacement. From each of the units selected at the first stage, consider selecting the second-stage units independently. Note that each time a first-stage unit is selected, the second-stage sample is selected from it independently.

With Y_i denoting the total of the first-stage unit, let $r_i = Y_i/u_i$. For Y_i, consider an unbiased estimator $\hat{Y}_i$ obtained from the sample at the second stage, and let $\hat{r}_i = \hat{Y}_i/u_i$. Now, an estimator for Y is given by

$$\hat{Y}_{pps} = \sum_i^n \hat{r}_i/n. \qquad (8.20)$$

Since $E_2(\hat{r}_i) = r_i$, $E(\hat{Y}_{pps}) = Y$. Thus, $\hat{Y}_{pps}$ is unbiased for Y. As shown in Appendix A8, the variance of $\hat{Y}_{pps}$ and its estimator are given by

$$V(\hat{Y}_{pps}) = \sum_i^N u_i(r_i - Y)^2/n + (1/n)\sum_i^N V_2(\hat{Y}_i)/u_i. \qquad (8.21)$$

and

$$v(\hat{Y}_{pps}) = \sum_i^n (\hat{r}_i - \hat{Y}_{pps})^2/n(n-1). \qquad (8.22)$$

Exercises

8.1.   For a sample of three from the nine clusters of Table 7.1 and a subsample of three counties from each of the selected clusters, find the variance for estimating the employment per county. Compare this with the variance of the mean of a simple random sample of nine counties. Note that the means and the variances of the clusters are presented in Table 7.2, and S^2 for the employment equals 5057.79.

8.2. With the results of the two-stage sampling considered in Example 8.2 and presented in Table 8.1, (a) estimate the average amount of employment per county and find its standard error. (b) Use these figures to estimate the total employment and its standard error.

8.3. Consider a sample of three clusters at the first stage from the nine clusters of Table 7.3. As in Example 8.3, at the second stage consider samples of size three for the first two and the last two clusters, and samples of size two for the rest of the five clusters. Find the standard errors for estimating the total and average employment for the 54 counties.

8.4. With the samples drawn in Example 8.4 at the two stages presented in Table 8.2, find (a) estimates for the total and the average employment and compare them with the actual values. (b) Find the standard errors from the samples.

8.5. With the sample sizes in Exercise 8.3, (a) find the standard errors for estimating the total and average employment through ratio adjustment with the cluster sizes. (b) Compare these standard errors with those in Exercise 8.3.

8.6. With the sample observations in Table 8.1, estimate the proportion and total number of counties that have at least 2000 establishments. Find the standard errors of the estimates.

8.7. Using the sample observations in Table 8.2 for the unequal size clusters, estimate the total and proportion of counties that have at least 2000 establishments. Obtain the estimates with and without ratio adjustment for sizes. Find the standard errors of the estimates.

8.8. *Project.* Consider the first three, next three, and the remaining four states in Table T7 in the Appendix as three clusters. Select all the samples of size two from each of the three clusters. From each selected cluster, select all the samples of two states. Select the clusters and the states randomly without replacement. For the

total number of physicians, from each of the samples
selected above in two stages, find the sample estimate in
Section 8.5 and the ratio estimate in (8.8). From these
results, (a) find the expectation and variance of the sample estimate and show that they coincide with the actual
total Y and the variance in (8.6). (b) Find the expectation,
bias, and MSE of the ratio estimator, and compare this
exact MSE with the approximation in (8.8).

Appendix A8

Expected value and variance of the estimator for the mean

The expected value and variance of $\bar{\bar{y}}$ can be obtained from the general
results in Appendix A3. From these expressions,

$$E(\bar{\bar{y}}) = E_1 E_2(\bar{\bar{y}})$$

and

$$V(\bar{\bar{y}}) = V_1 E_2(\bar{\bar{y}}) + E_1 V_2(\bar{\bar{y}}),$$

where the subscript 2 refers to the expectation conditional on the units
selected at the first stage and 1 to the unconditional expectation.
   Since $\bar{y}_i$ is unbiased for the mean $\bar{Y}_i$ of the selected primary unit,

$$E_2(\bar{\bar{y}}) = E_2\left(\frac{1}{n}\sum_1^n \bar{y}_i\right) = \frac{1}{n}\sum_1^n \bar{Y}_i$$

and

$$E(\bar{\bar{y}}) = E_1 E_2(\bar{\bar{y}}) = \frac{1}{N}\sum_1^N \bar{Y}_i = \bar{\bar{Y}}.$$

Thus, $\bar{\bar{y}}$ is unbiased for $\bar{\bar{Y}}$.

Further,

$$V_1 E_2(\bar{\bar{y}}) = (1 - f_1)S_1^2/n,$$

where $f_1 = n/N$ is the sampling fraction at the first stage.

Samples at the second stage are drawn independently from the selected primary units. Consequently, the covariance between $\bar{y}_i$ and $\bar{y}_j$ for $(i \neq j)$ vanishes. Hence,

$$V_2(\bar{\bar{y}}) = V_2\left(\sum_1^n \bar{y}_i\right)\bigg/n^2 = [(1 - f_2)/m]\sum_1^n S_{2i}^2/n^2,$$

where $f_2 = m/M$ is the sampling fraction at the second stage. From this expression,

$$E_1 V_2(\bar{\bar{y}}) = (1 - f_2)S_2^2/nm.$$

Finally,

$$V(\bar{\bar{y}}) = V_1 E_2(\bar{\bar{y}}) + E_1 V_2(\bar{\bar{y}}),$$

which is presented in (8.4).

Note that

$$(n - 1)s_1^2 = \sum_1^n (\bar{y}_i - \bar{\bar{y}})^2 = \sum_1^n [(\bar{y}_i - \bar{\bar{Y}}) - (\bar{\bar{y}} - \bar{\bar{Y}})]^2$$

$$= \sum_1^n (\bar{y}_i - \bar{\bar{Y}})^2 - n(\bar{\bar{y}} - \bar{\bar{Y}})^2.$$

Hence,

$$(n - 1)E(s_1^2) = E\sum_1^n (\bar{y}_i - \bar{\bar{Y}})^2 - nV(\bar{\bar{y}}).$$

Now,

$$\sum_1^n (\bar{y}_i - \bar{\bar{Y}})^2 = \sum_1^n [(\bar{y}_i - \bar{Y}_i) + (\bar{Y}_i - \bar{\bar{Y}})]^2$$

$$= \sum_1^n (\bar{y}_i - \bar{Y}_i)^2 + \sum_1^n (\bar{Y}_i - \bar{\bar{Y}})^2 + 2\sum_1^n (\bar{y}_i - \bar{Y}_i)(\bar{Y}_i - \bar{\bar{Y}}).$$

Since $E_2(\bar{y}_i) = \bar{Y}_i$, expectation of the last term vanishes. Further, $E_2(\bar{y}_i - \bar{Y}_i)^2 = V_2(\bar{y}_i) = (1 - f_2)S_{2i}^2/m$, and hence $E\sum_1^n (\bar{y}_i - \bar{Y}_i)^2 = n[(1 - f_2)/m]\sum_1^N S_{2i}^2/N = n(1 - f_2)S_2^2/m$. Expectation of the second term is equal to $(n/N)\sum_1^N (\bar{Y}_i - \bar{\bar{Y}})^2 = n(N - 1)S_1^2/N$.

From these results,

$$(n - 1)E(s_1^2) = n(1 - f_2)S_2^2/m + n(N - 1)S_1^2/N - nV(\bar{\bar{y}})$$

$$= (n - 1)(1 - f_2)S_2^2/m + (n - 1)S_1^2.$$

Thus,

$$s_1^2 - (1 - f_2)s_2^2/m \text{ is unbiased for } S_1^2.$$

Expectation and variance of the estimator for the total

Since

$$E_2(\hat{Y}) = N\sum_1^n \hat{Y}_i/n$$

$$E(\hat{Y}) = E_1[E_2(\hat{Y})] = \sum_1^N Y_i = Y.$$

Hence, $\hat{Y}$ is unbiased for the population total. Now,

$$V_1[E_2(\hat{Y})] = N^2[(1 - f_1)/n]\sum_1^N (Y_i - \bar{Y})^2/(N - 1).$$

Noting that the covariance between $\hat{Y}_i$ and $\hat{Y}_j$ for $(i \neq j)$ vanishes, from (8.10),

$$V_2(\hat{Y}) = (N/n)^2 \sum_1^n M_i^2 (1 - f_{2i}) S_{2i}^2/m_i,$$

where $f_{2i} = m_i/M_i$. Hence,

$$E_1[V_2(\hat{Y})] = (N/n) \sum^N M_i^2 (1 - f_{2i}) S_{2i}^2/m_i.$$

Finally, $V(\hat{Y}) = V_1[E_2(\hat{Y})] + E_1[V_2(\hat{Y})]$, which is presented in (8.6).

Estimation of the variance — unequal cluster sizes

Note that

$$E\left[\sum_1^n (\hat{Y}_i - \hat{\bar{Y}})^2\right] = [(n - 1)/N] \sum_1^N M_i^2 (1 - f_{2i}) S_{2i}^2/m_i$$

$$+ [(n - 1)/(N - 1)]\left[\sum_1^N (Y_i - \bar{Y})^2\right].$$

Thus,

$$N^2[(1 - f_1)/n]E\left[\sum_1^n (\hat{Y}_i - \hat{\bar{Y}})^2/(n - 1)\right] =$$

$$N[(1 - f_1)/n]\sum_1^N M_i^2 (1 - f_{2i}) S_{2i}^2/m_i$$

$$+ N^2[(1 - f_1)/n]\sum_1^N (Y_i - \bar{Y})^2/(N - 1).$$

Further,

$$E\left[(1/n)\sum_1^n M_i^2 \frac{1-f_{2i}}{m_i} s_{2i}^2\right] = (1/N)\sum_1^N M_i^2 \frac{1-f_{2i}}{m_i} S_{2i}^2.$$

$$E\left[(1/n)\sum_1^n M_i^2(1-f_{2i})s_{2i}^2/m_i\right] = (1/N)\sum_1^N M_i^2(1-f_{2i})S_{2i}^2/m_i.$$

Variance of the ratio estimator for the total

From (8.8),

$$E_2(\hat{Y}_R) = N\bar{M}\left(\sum_1^n M_i(\bar{Y}_i/n)\right)/\bar{m}.$$

Hence, for large n,

$$V_1 E_2(\hat{Y}_R) = N^2[(1-f_1)/n]\sum^N M_i^2(\bar{Y}_i - \bar{\bar{Y}})^2/(N-1).$$

Now,

$$V_2(\hat{Y}_R) = (N/n)^2\left[\sum_i^n M_i^2(1-f_{2i})S_{2i}^2/m_i\right]$$

$$E_1 V_2(\hat{Y}_R) = (N/n)\left[\sum^N M_i^2(1-f_{2i})S_{2i}^2/m_i\right].$$

The variance of $\hat{Y}_R$ is obtained by combining the second and last terms. The conditional expectation of (8.17) is

$$E_2(\hat{Y}_{HT}) = \sum_1^n \frac{1}{\phi_i} E_2(\hat{Y}_i) = \sum_1^n \frac{Y_i}{\phi_i}$$

and hence

$$E(\hat{Y}_{HT}) = E_1\left(\sum_1^n \frac{Y_i}{\phi_i}\right) = \sum_1^N Y_i = Y.$$

Now, with the procedure in Section 7.12,

$$V_1[E_2(\hat{Y}_{HT})] = \sum_i^N \sum_{i<j}^N (\phi_i\phi_j - \phi_{ij})\left(\frac{Y_i}{\phi_i} - \frac{Y_j}{\phi_j}\right)^2.$$

Noting that $\hat{Y}_i$ and $\hat{Y}_j$ are uncorrelated, from (8.17),

$$V_2(\hat{Y}_{HT}) = \sum_1^n \frac{V_{2i}}{\phi_i^2}$$

and hence

$$E_1[V_2(\hat{Y}_{HT})] = \sum_1^N \frac{V_{2i}}{\phi_i}.$$

Finally,

$$V(\hat{Y}_{HT}) = V_1[E_2(\hat{Y}_{HT})] + E_1[V_2(\hat{Y}_{HT})]$$

$$= \sum_i^N \sum_{i<j}^N (\phi_i\phi_j - \phi_{ij})\left(\frac{Y_i}{\phi_i} - \frac{Y_j}{\phi_j}\right)^2 + \sum_1^N \frac{V_{2i}}{\phi_i}.$$

Estimator for $V(\hat{Y}_{HT})$

As suggested by Des Raj (1968, p. 118),

$$E_2\left(\frac{\hat{Y}_i}{\phi_i} - \frac{\hat{Y}_j}{\phi_j}\right)^2 = V_2\left(\frac{\hat{Y}_i}{\phi_i} - \frac{\hat{Y}_j}{\phi_j}\right) + \left(\frac{Y_i}{\phi_i} - \frac{Y_j}{\phi_j}\right)^2.$$

Since the samples at the second stage are selected independently, the covariance between $\hat{Y}_i$ and $\hat{Y}_j$ vanishes, and this equation can be expressed as

$$E_2\left(\frac{\hat{Y}_i}{\phi_i} - \frac{\hat{Y}_j}{\phi_j}\right)^2 = \left(\frac{V_{2i}}{\phi_i^2} + \frac{V_{2j}}{\phi_j^2}\right) + \left(\frac{Y_i}{\phi_i} - \frac{Y_j}{\phi_j}\right)^2.$$

Now,

$$E\left[\sum_i^n \sum_{i<j}^n \left(\frac{(\phi_i\phi_j - \phi_{ij})}{\phi_{ij}}\right)\left(\frac{\hat{Y}_i}{\phi_i} - \frac{\hat{Y}_j}{\phi_j}\right)^2\right]$$

$$= \sum_i^N \sum_{i<j}^N (\phi_i\phi_j - \phi_{ij})\left[\left(\frac{V_{2i}}{\phi_i^2} + \frac{V_{2j}}{\phi_j^2}\right) + \left(\frac{Y_i}{\phi_i} - \frac{Y_j}{\phi_j}\right)^2\right]$$

$$= \sum_i^N \sum_{i<j}^N (\phi_i\phi_j - \phi_{ij})\left(\frac{V_{2i}}{\phi_i^2} + \frac{V_{2j}}{\phi_j^2}\right) + \sum_i^N \sum_{i<j}^N (\phi_i\phi_j - \phi_{ij})\left(\frac{Y_i}{\phi_i} - \frac{Y_j}{\phi_j}\right)^2.$$

Following (7.33c), the first term of this expression is the same as $\sum_i^N (V_{2i}/\phi_i^2)\phi_i(1 - \phi_i) = \sum_i^N V_{2i}/\phi_i - \sum_i^N V_{2i}$. Thus,

$$E\left[\sum_i^n \sum_{i<j}^n \left(\frac{(\phi_i\phi_j - \phi_{ij})}{\phi_{ij}}\right)\left(\frac{\hat{Y}_i}{\phi_i} - \frac{\hat{Y}_j}{\phi_j}\right)^2\right] = V(\hat{Y}_{HT}) - \sum_i^N V_{2i}.$$

At the second stage, let $\hat{V}_{2i}$ denote an unbiased estimator for V_{2i}, that is, $E_2(\hat{V}_{2i}) = V_{2i}$. Now, $\sum_i^n (\hat{V}_{2i}/\phi_i)$ is unbiased for $\sum_i^N V_{2i}$, and from the above expression an unbiased estimator for $V(\hat{Y}_{HT})$ is given by

$$v(\hat{Y}_{HT}) = \sum_i^n \sum_{i<j}^n \left(\frac{\phi_i\phi_j - \phi_{ij}}{\phi_{ij}}\right)\left(\frac{\hat{Y}_i}{\phi_i} - \frac{\hat{Y}_j}{\phi_j}\right)^2 + \sum_1^n \frac{\hat{V}_{2i}}{\phi_i}.$$

Variance of $\hat{Y}_{pps}$ and its estimator

First note that since $\hat{Y}_i$ are independent, $V(\hat{Y}_{pps}) = V(\hat{r}_i)/n$. Now, as found in Section 7.13, $V_1 E_2(\hat{r}_i) = V_1(r_i) = \sum_i u_i(r_i - Y)^2$. Further, $V_2(\hat{r}_i) =$

$V_2(\hat{Y}_i)/u_i^2$, and $E_1 V_2(\hat{r}_i) = \sum_i^N V_2(\hat{Y}_i)/u_i$. Thus,

$$V(\hat{Y}_{pps}) = \sum_i^N u_i(r_i - Y)^2/n + (1/n)\sum_i^N V_2(\hat{Y}_i)/u_i.$$

As in section 7.13, an unbiased estimator of $V(\hat{r}_i)$ is given by $v(\hat{r}_i) = \sum_i^n (\hat{r}_i - \hat{Y}_{pps})^2/(n-1)$. Hence an unbiased estimator of $V(\hat{Y}_{pps})$ is given by $v(\hat{Y}_{pps}) = \sum_i^n (\hat{r}_i - \hat{Y}_{pps})^2/n(n-1)$.

CHAPTER 9

Ratios and Ratio Estimators

9.1 Introduction

For the math (y_i) and verbal (x_i) scores examined in Chapter 2, the ratio of the math to verbal score, $r_i = y_i/x_i$, or their ratios to the total score, $r_i = y_i/(x_i + y_i)$ and $r_i = x_i/(x_i + y_i)$, are frequently of importance. Similarly, the ratios of the employment and establishment figures examined in Chapters 7 and 8 are of practical interest. Per capita figures of agricultural productions, Gross National Products (GNPs), energy consumptions, and similar characteristics for each country are obtained by dividing their totals (y_i) by the population sizes (x_i). The average of these types of ratios $\bar{R} = \Sigma r_i/N$ can be estimated from the sample mean $\bar{r} = \Sigma r_i/n$ of the ratios. Just as the sample mean $\bar{y}$ is unbiased for the population mean $\bar{Y}$, this average of the sample ratios is unbiased for $\bar{R}$. Its variance and estimator of variance are easily obtained from (2.12) and (2.14) by replacing y_i with r_i.

In contrast to the above average of the ratios, the ratio of the means or totals of two characteristics is given by $R = \bar{Y}/\bar{X} = Y/X$. Ratio of the average of the math scores to the average of the total (math + verbal) scores is an example. Another illustration is provided by the ratio of the average of the employment figures to the average of the establishments of the type examined in Chapter 7. From the observations (x_i, y_i), $i = 1, 2, ..., n$ of a random sample selected without replacement from the N population units, $\hat{R} = \bar{y}/\bar{x}$ is an estimator for R. In single-stage as well as multistage sampling, several estimators take the form of the ratio of two means. Chapters 7 and 8 have shown that in the case of clusters of unequal size, the population total or mean can be estimated with high precision by this type of ratio, with x_i representing the cluster size.

In the above type of illustrations, x_i can also provide concomitant or supplementary information on y_i. If x_i and y_i are positively correlated, $\bar{Y}/\bar{X}$ can be expected to be close to the sample ratio $\bar{y}/\bar{x}$ and

$\bar{Y}$ can be estimated from $(\bar{y}/\bar{x})\bar{X}$ with high precision. Further gains in precision can be obtained by combining this procedure with stratification. P.S.R.S. Rao (1998a) presents a summary of the ratio estimation procedures.

9.2   Bias and variance of the sample ratio

The sample ratio $\hat{R}$ is biased for R even when x_i and y_i are uncorrelated. From the **linearization** method in Appendix A9, this bias becomes negligible for large n. In such a case, the MSE of $\hat{R}$, which becomes the same as its variance, is approximately given by

$$V(\hat{R}) = (1 - f)S_d^2/n\bar{X}^2 \qquad (9.1)$$

where

$$S_d^2 = \sum_1^N (y_i - Rx_i)^2/(N - 1)$$
$$= S_y^2 + R^2 S_x^2 - 2RS_{xy} = S_y^2 + R^2 S_x^2 - 2R\rho S_x S_y.$$

An estimator of S_d^2 is obtained from

$$s_d^2 = \sum_1^n (y_i - \hat{R}x_i)^2/(n - 1) = s_y^2 + \hat{R}^2 s_x^2 - 2\hat{R}s_{xy}.$$

Hence, one may estimate (9.1) from

$$v(\hat{R}) = (1 - f)s_d^2/n\bar{x}^2. \qquad (9.2)$$

Example 9.1. Heights and weights: For the sake of illustration, Table 9.1 presents heights, weights, and systolic and diastolic blood pressures of $N = 15$ candidates along with their means, variances, standard deviations, and correlations. Consider height (x_i) and weight (y_i). The population ratio of the averages of the weights and heights is $R = 155.53/67.73 = 2.3$. Further, $S_d^2 = 187.12 + (2.3)^2(8.49) - 2(2.3)(0.62)$ $(2.91)(13.68) = 118.5$. If one considers a random sample of $n = 5$, from (9.1), $V(\hat{R}) = 0.0034$ and hence S.E. $(\hat{R}) = 0.0581$. In a random sample of size five, units (4, 6, 9, 12, 15) were selected. From the observations of these units, the means and variances of heights and weights are $(\bar{x}, \bar{y}) =$

(66, 158.6) and $(s_x^2, s_y^2) = (24, 314.8)$. The estimate of R is $\hat{R} = 158.6/66 = 2.4$. Further, $s_{xy} = 84.5$ and the sample correlation coefficient is 0.97. With these figures, $s_d^2 = [314.8 + 5.76(24) - 2(2.4)(84.5)] = 47.44$. Now, from (9.2), $v(\hat{R}) = 0.0015$ and S.E.$(\hat{R}) = 0.0381$.

9.3   Confidence limits for the ratio of means

Approximate confidence limits for R can be obtained from $\hat{R} \pm Z\sqrt{v(\hat{R})}$. If (x_i, y_i), $i = 1, 2, ..., n$, are considered to be observations from a bivariate normal distribution, Z is replaced by the $(1 - \alpha)$ percentile of the t-distribution with $(n - 1)$ d.f.

Alternative limits for R may be obtained by adopting Fieller's (1932) approach. If (x, y) follow a bivariate normal distribution, $\bar{d} = (\bar{y} - R\bar{x})$ follows a normal distribution with mean zero and variance $V(\bar{d}) = V(\bar{y}) + R^2 V(\bar{x}) - 2R\text{Cov}(\bar{y}, \bar{x})$. For sampling from a finite population, an unbiased estimator of this variance for a *given* R is $v(\bar{d}) = [(1 - f)/n](s_y^2 + R^2 s_x^2 - 2Rs_{xy})$. Now, confidence limits for R are obtained by expressing $t_{n-1}^2 = (\bar{y} - R\bar{x})^2/v(\bar{d})$ as a quadratic equation in R and finding its two roots.

Example 9.2. Confidence limits for the ratio of means: This example examines the limit for the ratio of the heights and weights of Example 9.1. For the probability of 0.95, $t_4 = 2.7764$. Since t_4 S.E.$(\hat{R}) = 1.06$, confidence limits for R are given by 2.4 ± 1.06, that is, (1.34, 3.46). With the normal approximation, Z S.E.$(\hat{R}) = 0.07$, and the limits are given by 2.4 ± 0.07, that is, (2.33, 2.47). The limits from the t-distribution are wider; that is, they are *conservative*.

For Fieller's method, $(2.7764)^2(2/15)[314.8 + 24R^2 - 2R(84.5)] = (158.6 - 66R)^2$. Simplifying this equation, $R^2 - 4.87R + 5.73 = 0$. The roots of this equation are (1.99, 2.88). In this example, these limits are also wider compared with the limits obtained above from the normal approximation.

9.4   Ratio estimators for the mean and total

If x_i and y_i are positively correlated and $\bar{X}$ is known, to estimate $\bar{Y}$, one may consider the ratio estimator:

$$\hat{\bar{Y}}_R = \frac{\bar{y}}{\bar{x}}\bar{X} = \hat{R}\bar{X}. \tag{9.3}$$

Table 9.1. Heights, weights, and blood pressures.

Unit	Height	Weight	Systolic Pressure	Diastolic Pressure
1	65	140	120	90
2	63	145	125	80
3	65	150	140	95
4	68	148	140	84
5	70	150	135	86
6	72	160	150	96
7	70	175	150	90
8	70	155	140	82
9	72	180	160	100
10	68	175	160	95
11	64	155	140	84
12	66	135	130	80
13	68	145	130	90
14	65	150	140	85
15	70	170	160	100
Mean	67.73	155.53	141.33	89.13
Variance	8.49	187.12	158.81	46.55
S.D.	2.91	13.68	12.60	6.82
Correlations				
	1			
	0.62	1		
	0.64	0.92	1	
	0.54	0.69	0.70	1

The bias of $\hat{\bar{Y}}_R$, $B(\hat{\bar{Y}}_R) = E(\hat{\bar{Y}}_R) - \bar{Y} = \bar{X}[E(\hat{R}) - R]$ becomes negligible for large n. The large sample MSE or variance of $\hat{\bar{Y}}_R$ is given by

$$V(\hat{\bar{Y}}_R) = E(\hat{\bar{Y}}_R - \bar{Y})^2 = \bar{X}^2 E(\hat{R} - R)^2$$
$$= \frac{(1-f)}{n} S_d^2 = \frac{(1-f)}{n}(S_y^2 + R^2 S_x^2 - 2R\rho S_x S_y). \quad (9.4)$$

From (2.12) and (9.4),

$$V(\bar{y}) - V(\hat{\bar{Y}}_R) = \frac{(1-f)}{n} R^2 S_x^2 \left(2\rho \frac{C_y}{C_x} - 1\right), \quad (9.5)$$

where $C_x = S_x/\bar{X}$ and $C_y = S_y/\bar{Y}$ are the coefficients of variation of the two characteristics. Hence, the ratio estimator has smaller variance than the sample mean if $\rho > (C_x/2C_y)$. Large positive correlation is helpful for the ratio estimator.

An estimator for (9.4) is given by

$$v(\hat{\bar{Y}}_R) = \frac{(1-f)}{n}s_d^2 = \frac{(1-f)}{n}(s_y^2 + \hat{R}^2 s_x^2 - 2\hat{R}s_{xy}). \qquad (9.6)$$

An alternative estimator is $v^1(\hat{\bar{Y}}_R) = \bar{X}^2 v(\hat{R}) = (\bar{X}/\bar{x})^2(1-f)s_d^2/n$.

The ratio estimator for the total Y is $\hat{Y}_R = N\hat{\bar{Y}}_R$ and its large sample variance is given by $V(\hat{Y}_R) = N^2 V(\hat{\bar{Y}}_R)$, which can be estimated from $v(\hat{Y}_R) = N^2 v(\hat{\bar{Y}}_R)$ or $V(\hat{Y}_R) = N^2 v^1/(\hat{\bar{Y}}_R)$.

Example 9.3. Gain in precision for the ratio estimator: For the illustration in Example 9.1 with $n = 5$, $V(\hat{\bar{Y}}_R) = (2/3)(118.5)/5 = 15.8$ and hence S.E.$(\hat{\bar{Y}}_R) = 3.97$. In this case, $V(\bar{y}) = (2/3)(187.12)/5 = 24.95$. Thus, the gain in precision for the ratio estimator relative to the sample mean is $(24.95 - 15.8)/15.8 = 0.58$ or 58%.

9.5   Confidence limits for the mean and total

With the five observed units in the sample, $\hat{\bar{Y}}_R = 2.4(67.73) = 162.55$, $v(\hat{\bar{Y}}_R) = [(15 - 5)/(15 \times 5)](47.44) = 6.33$ and S.E.$(\hat{\bar{Y}}_R) = 2.52$. Now, the 95% confidence limits for $\bar{Y}$ are given by $162.55 - 1.96(2.52) = 157.61$ and $162.55 + 1.96(2.52) = 167.49$.

From the above sample of five units, $\hat{Y}_R = 15(162.55) = 2438.25$ and S.E.$(\hat{Y}_R) = 15(2.52) = 37.8$. Hence, the 95% confidence limits for Y are given by $2438.25 - 1.96(37.8) = 2364.16$ and $2438.25 + 1.96(37.8) = 2512.34$. These limits can also be obtained by multiplying the above limits for the mean by the population size 15.

9.6   Differences between ratios, means, or totals

A common supplementary variable

Consider estimating the difference in the blood pressure–weight ratios for the systolic and diastolic measurements. With the subscripts 1 and 2 denoting the two types of measurements, these ratios are $R_1 = \bar{Y}_1/\bar{X}$ and $R_2 = \bar{Y}_2/\bar{X}$. From a sample of n units, $\hat{R}_1 = \bar{y}_1/\bar{x}$ and $\hat{R}_2 = \bar{y}_2/\bar{x}$ provide the estimators for these ratios. For large samples, the difference $\hat{R}_1 - \hat{R}_2$ is approximately unbiased for $R_1 - R_2$. With $d_{1i} = y_{1i} - \hat{R}_1 x_i$,

$d_{2i} = y_{2i} - \hat{R}_2 x_i$, and $s_d^2 = \Sigma_i(d_{1i} - d_{2i})^2/(n-1)$, an estimate of its variance is given by

$$v(\hat{R}_1 - \hat{R}_2) = \frac{(1-f)}{n\bar{x}^2} s_d^2. \tag{9.7}$$

The ratio estimators for $\bar{Y}_1$ and $\bar{Y}_2$ are given by $\hat{\bar{Y}}_{R1} = \hat{R}_1 \bar{X}$ and $\hat{\bar{Y}}_{R2} = \hat{R}_2 \bar{X}$, which are approximately unbiased. The difference $(\bar{Y}_1 - \bar{Y}_2)$ can now be estimated from $(\hat{\bar{Y}}_{R1} - \hat{\bar{Y}}_{R2})$, and an estimator for its large sample variance is provided by $(1-f)s_d^2/n$ with the above expression for s_d^2.

Example 9.4. Systolic and diastolic pressures: For the five sample units (4, 6, 9, 12, 15) considered in Example 9.1, the mean and variance of the systolic and diastolic pressures, respectively, are $(\bar{y}_1, s_1^2) = (148, 170)$ and $(\bar{y}_2, s_2^2) = (92, 88)$. The sample covariance of these two types of measurements is $s_{12} = 120$. The mean and variance of the weights are $(\bar{x}, s_x^2) = (158.6, 314.8)$.

Ignoring the information on the weights, an estimator for the difference of the means of the systolic and diastolic pressures is $(\bar{y}_1 - \bar{y}_2) = 56$. Following Section 3.5, $v(\bar{y}_1 - \bar{y}_2) = [(15 - 5)/(15 \times 5)](170 + 88 - 240) = 2.4$, and hence S.E.$(\bar{y}_1 - \bar{y}_2) = 1.55$.

From the sample, ratios of the means of the systolic and diastolic pressures to the mean of the weights are $\hat{R}_1 = 148/158.6 = 0.9332$ and $\hat{R}_2 = 92/158.6 = 0.5801$. Since $\bar{X} = 155.53$, the ratio estimators for the population means of the two types of pressures are $\hat{\bar{Y}}_{R1} = (0.9332)155.53 = 145.14$ and $\hat{\bar{Y}}_{R2} = (0.5801)155.53 = 90.22$. Hence, $\hat{\bar{Y}}_{R1} - \hat{\bar{Y}}_{R2} = 54.92$. From the sample observations, $s_d^2 = 9.5797$. Now, from (9.7), $v(\hat{\bar{Y}}_{R1} - \hat{\bar{Y}}_{R2}) = 1.28$ and S.E.$(\hat{\bar{Y}}_{R1} - \hat{\bar{Y}}_{R2}) = 1.13$.

The actual difference of the population means is $141.33 - 89.13 = 52.2$, and the estimates from the sample means as well as the ratio estimation are not too far from this figure.

From the sample means, an estimate for the difference of the population totals is $15(56) = 840$, and it has an S.E. of $15(1.55) = 23.25$. From the ratio method, an estimate for this difference is $15(54.92) = 819.3$ which has an S.E. of $15(1.13) = 16.95$.

Different supplementary variables

Denote the initial observations on the supplementary and main variables of a population of N units by (x_{1i}, y_{1i}) and their observations on another occasion by (x_{2i}, y_{2i}). From a random sample of size n from the

N units, the population ratios on the two occasions, $R_1 = \bar{Y}_1/\bar{X}_1$ and $R_2 = \bar{Y}_2/\bar{X}_2$, can be estimated from $\hat{R}_1 = \bar{y}_1/\bar{x}_1$ and $\hat{R}_2 = \bar{y}_2/\bar{x}_2$. With $d_{1i} = y_{1i} - \hat{R}_1 x_{1i}$ and $d_{2i} = y_{2i} - \hat{R}_2 x_{2i}$, the variance of $(\hat{R}_1 - \hat{R}_2)$ can be estimated from

$$v(\hat{R}_1 - \hat{R}_2) = \frac{(1-f)}{n} \sum_i (d_{1i}/\bar{x}_1 - d_{2i}/\bar{x}_2)^2/(n-1). \qquad (9.8)$$

The ratio estimators for $\bar{Y}_1$ and $\bar{Y}_2$ are now given by $\hat{\bar{Y}}_{R1} = \hat{R}_1\bar{X}_1$ and $\hat{\bar{Y}}_{R2} = \hat{R}_2\bar{X}_2$. The variance of $(\hat{\bar{Y}}_{R1} - \hat{\bar{Y}}_{R2})$ can be estimated from

$$v(\hat{\bar{Y}}_{R1} - \hat{\bar{Y}}_{R2}) = \frac{(1-f)}{n} \sum_i (d_{1i} - d_{2i})^2/(n-1). \qquad (9.9)$$

Example 9.5. Reduction in the systolic pressure after an exercise program: Represent the initial measurements of the 15 sample units in Table 9.1 by the subscript 1. The initial means and variances of the weights and systolic pressures for the five sample units (4, 6, 9, 12, 15), respectively, are (158.6, 314.8) and (148, 170).

One can represent the measurements after a physical fitness program by the subscript 2. With this notation, the population means of the weights and systolic pressures in this case can be denoted by $(\bar{X}_2, \bar{Y}_2)$. For the five sample units, as an illustration, consider $(x_{2i}, y_{2i}) = (140, 135)$, (152, 145), (170, 155), (130, 128), and (165, 158). From these observations, the sample means and variances of the weights and systolic pressures now are (151.4, 279.8) and (144.2, 163.7). The sample covariance of the systolic pressures on the two occasions is $s_{y12} = 165.5$.

An unbiased estimator of $(\bar{Y}_1 - \bar{Y}_2)$ is $(\bar{y}_1 - \bar{y}_2) = 148 - 144.2 = 3.8$, that is, the systolic pressure has decreased by 3.8 after the fitness program. The sample variance of this estimator is $v(\bar{y}_1 - \bar{y}_2) = (2/15)$ $[170 + 163.7 - 2(165.5)] = 0.36$, and hence S.E.$(\bar{y}_1 - \bar{y}_2) = 0.6$.

As before, $\hat{R}_1 = 148/158.6 = 0.9332$ and $\hat{\bar{Y}}_{R1} = 0.9332(155.53) = 145.14$. For the observations after the program, $\hat{R}_2 = 144.2/151.4 = 0.9524$. If $\bar{X}_2 = 150$, $\hat{\bar{Y}}_{R2} = 0.9524(150) = 142.86$. Now, the estimate of $(\bar{Y}_1 - \bar{Y}_2)$ is $145.14 - 142.86 = 2.28$. From (9.9), $v(\hat{\bar{Y}}_{R1} - \hat{\bar{Y}}_{R2}) = (2/15)(0.4183) = 0.0558$ and hence S.E.$(\hat{\bar{Y}}_{R1} - \hat{\bar{Y}}_{R2}) = 0.24$.

9.7 Regression through the origin and the BLUEs

This section examines the classical regression through the origin for **infinite** populations, and recognizes its role in providing motivation for ratio-type estimators.

Least squares

Regressions of agricultural yields on acreage, industrial productions
on employee sizes, household expenses on family sizes, and the like
can be represented by the model

$$y_i = \beta x_i + \varepsilon_i, \quad i = 1,...,n. \tag{9.10}$$

The expectation of the residual or error ε_i at a given x_i, $E(\varepsilon_i | x_i)$, is
assumed to be zero. For this **regression through the origin**, the
mean of y_i at x_i, $E(y_i | x_i)$ is given by βx_i; it becomes zero at $x_i = 0$.
The plot of $E(y_i | x_i)$ against x_i is a straight line going through the origin
$(x_i, y_i) = (0, 0)$. The coefficient β is the **slope** of this line, which is the
rate of change of $E(y_i | x_i)$ with respect to x_i.

For some applications, the variance of y_i at a given x_i, $V(y_i | x_i)$, which
is the same as the variance of ε_i at a given x_i, $V(\varepsilon_i | x_i)$, is found to be
proportional to x_i; that is, it is of the form $\sigma^2 x_i$. The variance of the
transformed variable $y_i/x_i^{1/2}$ becomes σ^2. The residuals ε_i and ε_j at x_i
and x_j are assumed to be uncorrelated. Now, minimization of
$\Sigma_1^n[(y_i - \beta x_i)^2/x_i]$ results in the least squares (LS) estimator
$\hat{\beta} = \Sigma y_i/\Sigma x_i = \bar{y}/\bar{x}$ for the slope β. This estimator is unbiased and its
variance is $V(\hat{\beta}) = \sigma^2/n\bar{x}$. Estimates of $E(y_i | x_i)$ and predictions of indi-
vidual y_i are obtained from $\hat{y}_i = \hat{\beta} x_i = (\bar{y}/\bar{x})x_i$.

An unbiased estimator of σ^2 is given by the residual or error mean
square, $\hat{\sigma}^2 = \Sigma_1^n[(y_i - \hat{\beta} x_i)^2/x_i]/(n-1)$, which has $(n-1)$ d.f. Now, the
variance of $\hat{\beta}$ can be estimated from $v(\hat{\beta}) = \hat{\sigma}^2/n\bar{x}$, and the S.E.$(\hat{\beta})$ is
obtained from the square root of this expression. The null hypothesis
that $\beta = 0$ against the alternative hypotheses that $\beta > 0$ or $\beta < 0$ can
be tested from the statistic $t = \hat{\beta}/S.E.(\hat{\beta})$, which follows Student's t-
distribution with $(n-1)$ d.f. If the null hypothesis is rejected, the
regression of y_i on x_i cannot be used for the estimation of $E(y_i | x_i)$ or
the prediction of y_i.

For the sake of illustration, consider the observations on height (x_i)
and weight (y_i) of the 15 candidates of Table 9.1. If the regression of
y_i on x_i is of the above type and the 15 candidates are considered to
constitute a random sample from a large population, the LS estimate
of the slope is $\hat{\beta} = \bar{y}/\bar{x} = 155.53/67.73 = 2.3$. The estimates $\hat{y}_i = \hat{\beta} x_i$
at $x_i = (63, 64, 65, 66, 68, 70, 72)$ are (144.9, 147.2, 149.5, 151.8, 156.4,
161, 165.6). The observations on the 15 units and the estimate of the
regression line are presented in Figure 9.1.

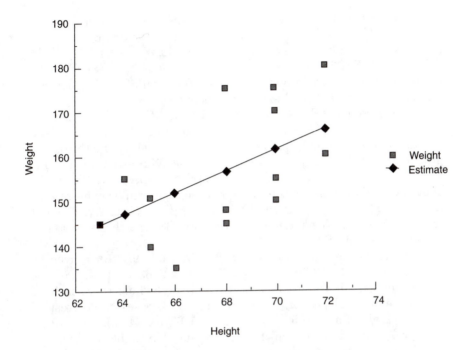

Figure 9.1. Regression through the origin of weight on height.

The estimate for σ^2 is 1.73. Thus, the variance of $\hat{\beta}$ is $1.73/67.73 = 0.0255$, and hence its S.E. is 0.1598. The value of $t = 2.3/0.1598 = 14.4$ with 14 d.f. is highly significant at the significance level of 0.001 or much smaller.

In general, the variance of ε_i can be of the form $\sigma^2 x_i^h = \sigma^2/W_i$, where $W_i = 1/x_i^h$. In several applications, the coefficient h has been found to be between zero and two. The approximate value of h can be found by plotting y_i against x_i and examining the variance of y_i at selected values of x_i. Now, for the generalized least squares (**GLS**) or weighted least squares (**WLS**) procedure, the slope is estimated by minimizing $\Sigma W_i (y_i - \beta x_i)^2$. As a result, $\hat{\beta} = \Sigma W_i x_i y_i / \Sigma W_i x_i^2$, which is unbiased with variance $V(\hat{\beta}) = \sigma^2/\Sigma W_i x_i^2$.

The LS estimator $\hat{\beta} x_i$ is unbiased for $E(y_i | x_i)$ and its variance is given by $x_i^2 V(\hat{\beta})$. This estimator is linear in the observations y_i, unbiased, and its variance is smaller than that of any other linear unbiased estimator; hence, it is known as the best linear unbiased estimator (**BLUE**).

An unbiased estimator of σ^2 is given by the residual or error mean square (EMS), $\hat{\sigma}^2 = \Sigma W_i(y_i - \hat{\beta}x_i)^2/(n-1)$. The error SS (ESS) can be expressed as $\Sigma W_i(y_i - \hat{\beta}x_i)^2 = \Sigma W_i y_i^2 + \hat{\beta}^2 \Sigma W_i x_i^2 - 2\hat{\beta}\Sigma W_i x_i y_i = \Sigma W_i y_i^2 - \hat{\beta}^2 \Sigma W_i x_i^2$ = total SS (TSS) − regression SS (RSS). For this regression through the origin, the d.f. for the total SS, regression SS, and error SS are n, 1, and $(n-1)$, respectively.

It has been noted that for some practical situations, $V(y_i \mid x_i)$ usually takes the form of σ^2, $\sigma^2 x_i$ or $\sigma^2 x_i^2$. For the first case, $V(y_i \mid x_i)$ is a constant; for the second, it is proportional to x_i; and for the third, it is proportional to x_i^2. For these three cases, the WLS estimator for β is given by $\Sigma x_i y_i/\Sigma x_i^2$, $\Sigma y_i/x_i = \bar{y}/\bar{x}$ and $\Sigma(y_i/x_i)/n$, respectively. From the above general expression, the ESS for these cases can be expressed as $\Sigma y_i^2 - \hat{\beta}^2 \Sigma x_i^2$, $\Sigma y_i^2/x_i - \hat{\beta}^2\Sigma x_i$, and $\Sigma(y_i/x_i)^2 - n\hat{\beta}^2$, respectively. The first term in each of these expressions is the total SS, and the second term is the regression SS.

For the data on heights and weights in Table 9.1, $\hat{\beta}$ is given by $10557.93/4595.73 = 2.2973$, $155.53/67.73 = 2.2963$, and $34.428/15 = 2.2952$ for the above three cases, respectively. Thus, there is little difference in the slopes for these cases. For these cases, the regression and residual SS are (1652.55, 363826), (5357.18, 24.16), and (79.02, 0.35), respectively. The ratio of the regression MS to the residual MS, which follows the F-distribution with 1 and $(n-1) = 15$ d.f. is highly significant for all three cases. From this result, the slope is inferred to be significantly greater than zero for all three cases.

The above regression approach provides motivation for estimating $\bar{Y}$ from $\bar{X}\Sigma W_i x_i y_i/\Sigma W_i x_i^2$. In particular, this estimator becomes the same as $\bar{Y}_R = (\bar{y}/\bar{x})\bar{X}$ in (9.3) when $V(y_i \mid x_i) = \sigma^2 x_i$. Note that, for the regression approach, the means and variances of y_i conditional on x_i are considered. The bias, variance, and MSE of the ratio estimators derived in the previous sections depend on the means and variances of both x and y, as well as their correlation.

Superpopulation model for evaluations and estimation

Assuming that the finite population of N units is a sample from an infinite **superpopulation**, following the model $y_i = \alpha + \beta x_i + \varepsilon_i$ with $V(y_i \mid x_i) = \sigma^2 x_i^h$, P.S.R.S. Rao (1968) compared the expected values of the MSE of $\hat{\bar{Y}}_R$ and the variances of $\hat{\bar{Y}}_{SS}$ and $\hat{\bar{Y}}_{RHC}$ described in Section 7.13; the cluster size is represented by x_i. This investigation showed that for $\alpha = 0$, that is, for the model in (9.10), $\hat{\bar{Y}}_R$ and $\hat{\bar{Y}}_{SS}$ are preferable when $V(y_i \mid x_i)$ is proportional to x_i and $\hat{\bar{Y}}_{RHC}$ when this variance is proportional to x_i^2.

When the finite population is assumed to be a sample from a superpopulation following the model in (9.10) with $V(y_i|x_i) = \sigma^2 x_i$, Royall (1970) considers $\hat{Y} = \Sigma_1^n c_i y_i$ to estimate Y. He shows that when the variance of $\hat{Y}$, $V(\hat{Y}) = E(\hat{Y} - Y)^2$, is minimized with the **model-unbiasedness** condition, $E(\hat{Y}) = E(Y)$, it becomes the same as the ratio estimator $\hat{Y}_R = \hat{R}X = (\bar{y}/\bar{x})X$. The variance of $\hat{Y}_R$ now becomes $V(\hat{Y}_R) = \sigma^2(N\bar{X} - n\bar{x}) N\bar{X}/n\bar{x}$. The estimator for σ^2 is given by $\hat{\sigma}^2 = \Sigma_1^n [(y_i - \hat{R}x_i)^2/x_i]/(n-1)$, which is the same as the least squares estimator presented earlier in this section. Cochran (1977, p. 159) also presents the derivations for $\hat{Y}_R$, $V(\hat{Y}_R)$, and $\hat{\sigma}^2$.

For the above approach, the estimator for $\bar{Y}$ is the same as $(\bar{y}/\bar{x})\bar{X}$ and its variance is estimated from $\hat{\sigma}^2(N\bar{X} - n\bar{x}) \bar{X}/Nn\bar{x}$. Through the model in (9.10), P.S.R.S. Rao (1981b) compares this variance estimator with $v(\hat{\bar{Y}}_R)$ in (9.6).

9.8 Ratio estimation vs. stratification

If the available supplementary variable is of the qualitative or categorical type, low–medium–high, for example, it can be used for stratifying the population, but not for the ratio estimation. However, if it is of the continuous type, ratio estimation is preferred provided the assumptions described in Section 9.7 are satisfied at least approximately.

To examine the two procedures, consider wheat production in 1997 (y) for the first $N = 20$ countries in Table T8 in the Appendix; the five countries with the largest production are not included in this group. Summary figures for these data are presented in Table 9.2.

If a sample of $n = 8$ countries is considered for the estimation of the average production, $V(\bar{y}) = [(20 - 8)/160](55.09) = 4.13$ and hence S.E.$(\bar{y}) = 2.03$. If one considers the supplementary data of 1995 (x), $R = 9.95/9.18 = 1.0839$ and $S_d^2 = 2.84$. Now, from (9.4), $V(\hat{Y}_R) = [(20 - 8)/160](2.84) = 0.213$ and S.E.$(\hat{Y}_R) = 0.46$. Thus, the S.E. of the ratio estimator is only about 20% of that of the sample mean.

To examine the effect of stratification, consider $G = 2$ strata, with the first $N_1 = 11$ and the next $N_2 = 9$ countries of Table T8 in the Appendix along with the summary figures in Table 9.2. For both proportional and Neyman allocations of the sample of $n = 8$ units for the two strata, $n_1 = 4$ and $n_2 = 4$ approximately. For these sample sizes, the variances of the sample means in the two strata are $V(\hat{Y}_1) = [(11 - 4)/44](8.14) = 1.3$ and $V(\hat{Y}_2) = [(9 - 4)/36](15.94) = 2.21$. Now, from (5.13), $V(\hat{Y}_{st}) = (11/20)^2(1.3) + (9/20)^2(2.21) = 0.84$ and hence

Table 9.2. Wheat production.

	1990	1995	1997
20 Countries			
Total	212.1	183.5	198.9
Mean	10.61	9.18	9.95
Variance	82.68	50.88	55.09
S.D.	9.09	7.13	7.42
11 Countries			
Total	53.1	41.2	45
Mean	4.83	3.75	4.09
Variance	21.29	8.08	8.14
S.D.	4.61	2.84	2.85
9 Countries			
Total	159	142.3	153.9
Mean	17.67	15.81	17.1
Variance	67.74	20.66	15.94
S.D.	8.23	4.55	3.99

	Correlations		
	(1990, 1995)	(1990, 1997)	(1995, 1997)
20 countries	0.89	0.90	0.98
11 countries	0.78	0.90	0.96
9 countries	0.77	0.81	0.88

Note: Productions for these 20 countries are presented in Table T8 in the Appendix.

S.E.$(\hat{\bar{Y}}_{st}) = 0.92$. Although this S.E. for stratification is only about 45% that of the sample mean, it is twice that of the ratio estimator.

In this illustration, the ratio estimator has much smaller S.E. than the sample mean since the correlation of the 1995 and 1997 production is 0.98, which is very high. The S.E. of the stratified estimator will be reduced if the 20 countries are divided into two strata with smaller within variances.

9.9   Ratio estimation with stratification

Since stratification as well as ratio estimation produce gains in precision, one may combine both procedures for further gains. There are two methods of achieving this objective, but first consider the benefits of ratio estimation in a single stratum.

Single stratum

When the mean of the supplementary variable $\bar{X}_g$ is known, the ratio estimators for the total Y_g and mean $\bar{Y}_g$ are

$$\hat{Y}_{Rg} = \frac{y_g}{x_g} X_g = \frac{\bar{y}_g}{\bar{x}_g} X_g \tag{9.11}$$

and

$$\hat{\bar{Y}}_{Rg} = \frac{\bar{y}_g}{\bar{x}_g} \bar{X}_g. \tag{9.12}$$

If n_g is large, the bias of (9.12) becomes negligible and following (9.4) its variance is approximately given by

$$\begin{aligned}
V(\hat{\bar{Y}}_{Rg}) &= (1 - f_g) S_{dg}^2 / n_g \\
&= \frac{(1 - f_g)}{n_g} [S_{yg}^2 + R_g^2 S_{xg}^2 - 2\rho_g S_{yg} S_{xg}],
\end{aligned} \tag{9.13}$$

where $R_g = \bar{Y}_g / \bar{X}_g$. Note that $S_{dg}^2 = \Sigma_1^{N_g} (y_{gi} - R_g x_{gi})^2 / (N_g - 1)$ and ρ_g is the correlation between x and y in the gth stratum. An estimator for this variance is

$$v(\hat{\bar{Y}}_{Rg}) = \frac{(1 - f_g)}{n_g} s_{dg}^2 = \frac{(1 - f_g)}{n_g} [s_{yg}^2 + \hat{R}_g^2 s_{xg}^2 - 2\hat{R}_g s_{xyg}], \tag{9.14}$$

where $\hat{R}_g = \bar{y}_g / \bar{x}_g$ and $s_{dg}^2 = \Sigma_1^{n_g} (y_{gi} - \hat{R}_g x_{gi})^2 / (n_g - 1)$.

The separate ratio estimator

If the regression of y on x approximately passes through the origin in each of the strata, one can consider the estimator in (9.11) separately for each of the strata totals. Now, adding these estimators, the ratio estimator for the population total $Y = \Sigma Y_g$ is given by

$$\hat{Y}_{RS} = \sum_1^G \hat{Y}_{Rg} = \sum_1^G \frac{y_g}{x_g} X_g = \sum_1^G N_g \frac{\bar{y}_g}{\bar{x}_g} \bar{X}_g \tag{9.15}$$

The subscripts R and S refer to ratio estimation and stratification. Dividing (9.15) by N, the estimator for $\bar{Y}$ is given by

$$\hat{\bar{Y}}_{RS} = \sum_1^G W_g \hat{\bar{Y}}_{Rg} = \sum_1^G W_g \frac{\bar{y}_g}{\bar{x}_g} \bar{X}_g \tag{9.16}$$

The variance and estimator of variance of (9.16) are

$$V(\hat{\bar{Y}}_{RS}) = \sum W_g^2 V(\hat{\bar{Y}}_{Rg}) \tag{9.17}$$

and

$$v(\hat{\bar{Y}}_{RS}) = \sum W_g^2 v(\hat{\bar{Y}}_{Rg}). \tag{9.18}$$

With proportional allocation of the sample, the variance in (9.17) becomes

$$V_{\text{prop}}(\hat{\bar{Y}}_{RS}) = \frac{(1-f)}{n} \sum_1^G W_g S_{dg}^2. \tag{9.19}$$

The variance in (9.17) or (9.19) clearly becomes small if y_{gi} is highly positively correlated with x_{gi} within each stratum. If these correlations are high and R_g vary, (9.19) can be expected to be much smaller than the variance of $\bar{Y}_R$ in (9.4).

For Neyman allocation, the variance in (9.17) is minimized for a given $n = \Sigma n_g$. As a result, n_g is chosen proportional to $N_g S_{dg}$. If the cost function in (5.33) is suitable, optimum allocation results in choosing n_g proportional to $N_g S_{dg}/\sqrt{e_g}$. These procedures require additional information.

Example 9.6. Wheat production: Consider estimation of the average wheat production for the 20 countries for 1997 (y) with the information from 1995 (x) through stratification and the ratio method. From the summary figures in Table 9.2, $S_{d1}^2 = 0.85$ and $S_{d2}^2 = 5.68$. For samples of size four from each of the strata, $V(\hat{\bar{Y}}_{R1}) = (7/44)(0.85) = 0.1352$ and $V(\hat{\bar{Y}}_{R2}) = (5/36)(5.68) = 0.7889$. Now, from (9.17), $V(\hat{\bar{Y}}_{RS}) = (11/20)^2(0.1352) + (5/36)^2(0.7889) = 0.20$ and S.E.$(\hat{\bar{Y}}_{RS}) = 0.45$. This variance is only one fourth of the variance of 0.84 found in Section 9.8 for the stratified mean and only slightly smaller than the variance of 0.213 for the ratio estimator without stratification.

The combined ratio estimator

If the ratios R_g for the strata do not differ much, as suggested by Hansen et al. (1946), it is possible first to obtain the common ratio as $\hat{R}_C = \hat{\bar{Y}}_{RC}/\hat{\bar{X}}_{RC} = \Sigma W_g \bar{y}_g / \Sigma W_g \bar{x}_g$ and for $\bar{Y}$ consider the **combined ratio estimator**:

$$\hat{\bar{Y}}_{RC} = \hat{R}_C \bar{X}. \tag{9.20}$$

For large samples, the bias of this estimator vanishes and its variance approximately becomes

$$V(\hat{\bar{Y}}_{RC}) = E(\hat{\bar{Y}}_{RC} - R\hat{\bar{X}}_{RC})^2$$

$$= \Sigma W_g^2 \frac{(1-f_g)}{n_g}(S_{yg}^2 + R^2 S_{xg}^2 - 2R\rho_g S_{yg} S_{xg}). \tag{9.21}$$

This variance can be estimated from

$$v(\hat{\bar{Y}}_{RC}) = \Sigma W_g^2 \frac{(1-f_g)}{n_g}(s_{yg}^2 + \hat{R}_c^2 s_{xg}^2 - 2\hat{R}_c s_{xyg}). \tag{9.22}$$

Derivations of (9.21) and (9.22) are outlined in Appendix A9.

From the figures in Table 9.2, $R = 9.95/9.18 = 1.084$. Now from (9.21), $V(\hat{\bar{Y}}_{RC}) = 0.1936$ and S.E.$(\hat{\bar{Y}}_{RC}) = 0.44$. This S.E. is almost the same as that of the separate estimator. One reason for this result is that the ratios of the means for the first and second strata are $R_1 = 4.09/3.75 = 1.09$ and $R_2 = 17.1/15.81 = 1.08$, which are almost the same.

Statistical test for the equality of the slopes

From samples of sizes n_g selected from large-size strata, the null hypothesis that their slopes are all the same against the alternative that they are unequal can be tested as follows.

1. Fit the G regressions through the origin, obtain the pooled residual sum of squares, SS$_{sep}$, with $(n_1 - 1) + (n_2 - 1) + \cdots + (n_G - 1) = (n - G)$ d.f. and obtain the estimate of σ^2 from $\hat{\sigma}^2 = SS_{sep}/(n - G)$.
2. Fit the regression through the origin from combining all the n observations and find the residual sum of squares,

SS_{comb}, with $(n - 1)$ d.f. The difference, $SS_{diff} = (SS_{comb} - SS_{sep})$ has $(n - 1) - (n - G) = (G - 1)$ d.f.

3.   Let $MS_{diff} = SS_{diff}/(G - 1)$. The ratio $F = MS_{diff}/\hat{\sigma}^2$ follows the F-distribution with $(G - 1)$ and $(n - G)$ d.f.

Percentiles of the F-distribution are tabulated and are also available through computer software packages. For a specified significance level, the null hypothesis is rejected if the computed value of F exceeds the actual percentage point. Rejection of the null hypothesis suggests the separate estimator; the combined estimator otherwise.

9.10   Bias reduction

From Appendix A9, ignoring terms of order $(1/n^2)$ and smaller, an estimate of the bias of $\hat{R}$ is given by

$$\hat{B}(\hat{R}) = (1 - f)\hat{R}(c_{xx} - c_{xy})/n, \qquad (9.23)$$

where $c_{xx} = s_x^2/\bar{x}^2$ and $c_{xy} = s_{xy}/\bar{x}\,\bar{y}$. Subtracting this expression from $\hat{R}$, an estimator for R is obtained from

$$\hat{R}_T = \hat{R}[1 - (1 - f)(c_{xx} - c_{xy})/n]. \qquad (9.24)$$

The bias of this estimator is of order $1/n^2$. The corresponding estimator for $\bar{Y}$ is given by $\hat{\bar{Y}}_T = \bar{X}\hat{R}_T$, which was suggested by Tin (1965).

The approach in Appendix A9 for expressing a nonlinear estimator such as $\hat{R}$ in a series and ignoring higher-order terms is known as the **linearization** procedure. As seen above, the bias of an estimator can be reduced by this approach. Further, as shown in Appendix A9, an approximation to the MSE of $\hat{R}$ and its estimator can also be obtained by this procedure. Chapter 12 compares this approach with the **jackknife** and **bootstrap procedures**.

Hartley and Ross (1954) derive an unbiased estimator for $\bar{Y}$ by removing the bias from $\bar{X}\bar{r}$. The investigation by P.S.R.S. Rao (1969) through the model in (9.10) showed that $\hat{\bar{Y}}_R$ and the Hartley–Ross estimators have relatively smaller MSEs when $V(y_i|x_i)$ is proportional to x_i and x_i^2, respectively.

9.11 Two-phase or double sampling ratio estimators

If $\bar{X}$ is not known, it can be estimated from the mean $\bar{x}_1$ of a large sample of size n_1 selected from the N population units. Now, $(\bar{x}, \bar{y})$ are obtained from a subsample of size n selected from the n_1 units. As an illustration, the first sample may provide an estimate for the average family size for the households in a region. The subsample at the second phase provides the averages for the family size and a major characteristic of interest, for example, savings. This type of two-phase sampling is feasible if the cost of sampling the first sample is cheaper than the second.

$\bar{Y}$ can now be estimated from

$$\hat{\bar{Y}}_{\mathrm{Rd}} = (\bar{y}/\bar{x})\bar{x}_1. \tag{9.25}$$

Following Appendix A3, for large n, the bias of this estimator becomes small and its variance becomes

$$V(\hat{\bar{Y}}_{\mathrm{Rd}}) = (N - n_1)S_y^2/Nn_1 + (n_1 - n)S_d^2/n_1 n. \tag{9.26}$$

Note that S_d^2 has the same expression as in Section 9.2.

The bias of $\hat{\bar{Y}}_{\mathrm{Rd}}$ can be reduced or eliminated through modifying and extending the procedures in Section 9.10. P.S.R.S. Rao (1981a) evaluates the efficiencies of nine such procedures through the model in (9.10). A summary of the double-sampling ratio estimators is presented in P.S.R.S. Rao (1998b).

In some situations, the means $\bar{x}_1$ and $(\bar{x}, \bar{y})$ are obtained from samples of sizes n_1 and n selected independently from the N population units. In such a case, one can consider the combined estimator $\bar{x}_a = a\bar{x}_1 + (1 - a)\bar{x}$ for $\bar{X}$. This estimator is unbiased and the coefficient a is obtained by minimizing its variance. As a result, the optimum value of a is given by $(N - n)n_1/[(N - n)n_1 + (N - n_1)n]$. Alternatively, one may remove the duplications from the two samples and consider the mean $\bar{x}_v$ of the v distinct units to estimate $\bar{X}$. The variances of both these alternative estimators are smaller than the variance of $\bar{x}$. Now, one can consider $(\bar{y}/\bar{x})\bar{x}_a$ or $(\bar{y}/\bar{x})\bar{x}_v$ to estimate $\bar{Y}$. P.S.R.S. Rao (1975b) evaluates the merits of these two estimators relative to $\hat{\bar{Y}}_{\mathrm{Rd}}$. Similar procedures for the Hartley–Ross type of estimation are examined in P.S.R.S. Rao (1975a).

9.12   Ratio estimator with unequal probability selection

As we have seen in Section 7.13 for the PPSS procedure, $\overline{X}(\overline{y}/\overline{x})$ is unbiased for $\overline{Y}$ if the sample (x_i, y_i), $i = 1, 2,..., n$, is selected with probability proportional to Σx_i.

If the units are selected with probabilities ϕ_i and without replacement, as described in Section 7.12 for the Horvitz–Thompson estimator, $\Sigma_1^n(y_i/\phi_i)$ and $\Sigma_1^n(x_i/\phi_i)$ are unbiased for $Y = \Sigma_1^N y_i$ and $X = \Sigma_1^N x_i$, but their ratio is not unbiased for R. For $\overline{Y}$, one can still consider the ratio-type estimator $[\Sigma_1^n(y_i/\phi_i)/\Sigma_1^n(x_i/\phi_i)]\overline{X}$. P.S.R.S. Rao (1991) empirically compared this procedure with the PPSS and pps procedures in Section 7.13 and the ratio estimator $\overline{X}(\overline{y}/\overline{x})$ with equal probability selection, with the sample size $n = 2$. The investigation showed that the MSE for this procedure can be smaller than that of the ratio estimator with equal probability selection and the variances of the estimators of the PPSS and pps procedures provided ϕ_i is proportional to y_i.

9.13   Multivariate ratio estimator

If the population means $\overline{X}_1, \overline{X}_2,..., \overline{X}_p$ of $p(>1)$ supplementary characteristics are available, the ratio estimators $\hat{\overline{Y}}_{R1} = \overline{X}_1(\overline{y}/\overline{x}_1)$, $\hat{\overline{Y}}_{R2} = \overline{X}_2(\overline{y}/\overline{x}_2),..., \hat{\overline{Y}}_{R2} = \overline{X}_2(\overline{y}/\overline{x}_2)$ can be considered for $\overline{Y}$. The means $(\overline{y}; \overline{x}_1, \overline{x}_2, ..., \overline{x}_p)$ are obtained from a sample of size n. The estimator corresponding to the supplementary variable that has the highest correlation with the principal characteristic (y) can be expected to have the smallest MSE. For $\overline{Y}$, Olkin (1958) considers the **multivariate** estimator $W_1\hat{\overline{Y}}_{R1} + W_2\hat{\overline{Y}}_{R2} + \cdots + W_p\hat{\overline{Y}}_{Rp}$, $\Sigma_1^p W_i = 1$. The weights $(W_1, W_2,..., W_p)$ are obtained by minimizing the MSE of this estimator.

Exercises

9.1.   *Project.* As in Exercise 2.10, consider the 20 possible samples of size three from the six candidates in Table 2.1. (a) Find the expectation of $\hat{R} = \overline{y}/\overline{x}$ and find its bias for estimating $R = \overline{Y}/\overline{X} = Y/X$. (b) Compare the exact variance and MSE of $\hat{R}$ with the approximate variance in (9.1). (c) Find the bias of (9.2) for estimating the approximate variance in (9.1) and the exact MSE of $\hat{R}$.

9.2. *Project.* With the results in Exercise 9.1, (a) find the bias of $\hat{\bar{Y}}_R$ for estimating $\bar{Y}$, (b) compare its approximate variance in (9.4) with its exact MSE, and (c) find the biases of $v(\hat{\bar{Y}}_R)$ and $v'(\hat{\bar{Y}}_R)$ for estimation of $V(\hat{\bar{Y}}_R)$.

9.3. *Project.* From the 20 samples in Exercise 9.1, (a) find the expectation of $\hat{R} = \bar{y}/(\bar{x} + \bar{y})$ and its bias in estimating $R = \bar{Y}/(\bar{X} + \bar{Y}) = Y/(X + Y)$. (b) Find the MSE of $\hat{R}$ and compare it with its approximate variance.

9.4. (a) From Table 9.1, find the population average $\bar{R}$ and variance $V(\bar{r})$ for the ratios of the systolic pressure (y) to weight (x). (b) From the sample units (4, 6, 9, 12, 15) selected in Example 9.1, estimate $\bar{R}$, find the sample variance of the estimate and the 95% confidence limits for $\bar{R}$.

9.5. From the five sample units (4, 6, 9, 12, 15) selected in Example 9.1, estimate $R = \bar{Y}/(\bar{X} + \bar{Y})$, find the S.E. of the estimate and the 95% confidence limits for this ratio.

9.6. With the sample units (3, 6, 8, 12, 13) selected from the 15 units of Table 9.1, (a) estimate the ratio $R = \bar{Y}/\bar{X}$ of the systolic pressure to weight and find the S.E. of the estimate, and (b) find the ratio estimator for $\bar{Y}$ and estimate its S.E.

9.7. With the sample units (3, 6, 8, 12, 13) selected from the 15 units of Table 9.1, (a) estimate the difference in the ratio estimates for the averages of the systolic and diastolic pressures with weight as the supplementary variable, (b) estimate the S.E., and (c) find the 95% confidence limits for the difference.

9.8. For the five sample units (4, 6, 9, 12, 15) selected in Example 9.1, consider the diastolic pressures (80, 92, 95, 80, 93) after the fitness program. Find the ratio estimate for the difference in the averages for the diastolic pressures before and after the fitness program and find the S.E. of the estimate.

9.9. From the data in Table 7.1, for a sample n clusters, (a) find the variances of the ratio estimators for the average employment per cluster and per county with the number of establishments as the supplementary variable. (b) Find the precisions of these estimators relative to the sample means.

9.10. To estimate the average employment of the unequal size clusters in Table 7.3, as seen in Example 7.5, the ratio estimator $(\bar{y}/\bar{m})\bar{M}$ has considerably higher precision than $\bar{y}$. An alternative ratio estimator with the number of establishments as the supplementary variable is $(\bar{y}/\bar{x})\bar{X}$. For another procedure, let $u_i = x_i/M_i$ and $v_i = y_i/M_i$, denote their sample means by $\bar{u}$ and $\bar{v}$, and consider the ratio estimator $(\bar{v}/\bar{u})\bar{X}$. For a sample of n clusters, find the precision of the three ratio estimators relative to the sample mean.

9.11.   For the population of the ten states with the highest enrollments presented in Table T4 in the Appendix, denote the public and private enrollments for 1990 by (x_1, y_1) and for 1995 by (x_2, y_2). (a) Find the means, standard deviations, covariances, and correlations of these four characteristics. (b) For both the years, with the total enrollment as the supplementary variable, find the variance of the ratio estimator for the mean of the public enrollments for a sample of four states. (c) Compare the variances in (b) with those of the sample means.

9.12.   (a) Find the variance of the difference of the ratio estimators in Exercise 9.11(b) for 1990 and 1995 and (b) compare it with the variance of the difference of the sample means.

9.13.   For the population of the ten states with the smallest enrollments presented in Table T5 in the Appendix, denote the public and private enrollments for 1990 by (x_1, y_1) and for 1995 by (x_2, y_2). (a) Find the means, standard deviations, covariances, and correlations of these four characteristics. (b) For both years, for a sample of four units, find the variance of the ratio of the totals of the public and private enrollments. (c) Find the variance of the difference of the ratios in (b).

9.14.   Consider the first six and the remaining nine units of Table 9.1 as two strata. With weight as the supplementary variable, compare the variances of the separate and combined ratio estimators for the mean of the systolic pressures for a total sample of size five. Consider both proportional and Neyman allocations.

9.15.   *Project.* For each of the 120 samples of size three that can be selected from the ten states in Table T7 in the Appendix, find the difference of the ratios of the means of the two types of expenditures to the total expenditure as in Section 9.6. (a) From the average and variance of the 120 estimates, find the bias and MSE of the above estimator. (b) Compute the average of the approximate estimates in (9.7) for each of the 120 samples, and compare their average with the exact MSE in (a).

9.16.   Consider the observations in Table 9.1 to be obtained from a sample of 15 units of a large population, and plot the observations on systolic pressure (y_i) against weight (x_i). As described in Section 9.7, consider regression through the origin of y_i on x_i when $V(y_i|x_i)$ equals σ^2, $\sigma^2 x_i$, and $\sigma^2 x_i^2$. For each of these three cases, estimate (a) the slope, (b) σ^2, and (c) the S.E. of the slope. (d) Test the hypothesis that $\beta = 0$ against the alternative hypothesis that $\beta > 0$. (e) From the above results, which of the three procedures do you recommend for estimating the mean of the systolic pressure; give reasons.

9.17.   Divide the 25 countries in Table T9 in the Appendix into two strata consisting of the eight largest petroleum-producing countries and the remaining 17 countries with smaller or no production. To estimate the average per capita energy consumption for the 25 countries with petroleum production as the supplementary variable, find the S.E. of the separate ratio estimator for samples of sizes four and eight from the two strata.

Appendix A9

Bias of the ratio of two sample means

The bias of $\hat{R} = \bar{y}/\bar{x}$ is given by

$$B(\hat{R}) = E(\hat{R}) - R = E[(\bar{y} - R\bar{x})/\bar{x}] = E[(\bar{y} - R\bar{x})/\bar{X}(1 + \delta)],$$

where $\delta = (\bar{x} - \bar{X})/\bar{X}$. Expressing $(1 + \delta)^{-1}$ as $1 - \delta + \delta^2 - \cdots$ and retaining only the first two terms,

$$B(\hat{R}) = E[(\bar{y} - R\bar{x})(1 - \delta)/\bar{X}]$$
$$= E(\bar{y} - R\bar{x})/\bar{X} - E[(\bar{y} - R\bar{x})(\bar{x} - \bar{X})]/\bar{X}^2.$$

The first term vanishes, and the bias now becomes

$$B(\hat{R}) = -[\text{Cov}(\bar{x}, \bar{y}) - RV(\bar{x})]/\bar{X}^2 = (1 - f)R(C_{xx} - C_{xy})/n,$$

where $C_{xx} = S_x^2/\bar{X}^2$ and $C_{xy} = S_{xy}\bar{X}\bar{Y}$. This bias becomes negligible for large n.

The above procedure of expressing a nonlinear estimator such as the ratio in a series and ignoring higher-order terms is known as the **linearization method**.

MSE of the ratio of two sample means

From the definition,

$$\text{MSE}(\hat{R}) = E(\hat{R} - R)^2 = E[(\bar{y} - R\bar{x})/\bar{x}]^2.$$

Following the linearization approach, for large n, the bias of $\hat{R}$ becomes negligible and this expression approximately becomes the variance of $\hat{R}$, given by

$$V(\hat{R}) = E(\bar{y} - R\bar{x})^2/\bar{X}^2.$$

To obtain an explicit expression for this equation, let $d_i = (y_i - Rx_i)$. The population mean of d_i is $\bar{D} = (\sum_1^N d_i)/N = \bar{Y} - R\bar{X} = 0$. Its population variance is

$$S_d^2 = \sum_1^N (y_i - Rx_i)^2/(N - 1) = \sum_1^N [(y_i - \bar{Y}) - R(x_i - \bar{X})]^2/(N - 1)$$
$$= S_y^2 + R^2 S_x^2 - 2RS_{xy} = S_y^2 + R^2 S_x^2 - 2R\rho S_x S_y.$$

The sample mean of the d_i is $\bar{d} = (\sum_1^n d_i)/n = (\bar{y} - R\bar{x})$. The expectation of $\bar{d}$ is zero and its variance is $E(\bar{d} - \bar{D})^2 = E(\bar{d}^2) = (1 - f)S_d^2/n$. Thus, from the above two expressions,

$$V(\hat{R}) = V(\bar{d})/\bar{X}^2 = (1 - f)S_d^2/n\bar{X}^2.$$

This expression can also be obtained by writing $E(\bar{y} - R\bar{x})^2$ as $E[(\bar{y} - \bar{Y}) - R(\bar{x} - \bar{X})]^2 = V(\bar{y}) + R^2 V(\bar{x}) - 2R \, \text{Cov}(\bar{x}, \bar{y})$.

From the sample,

$$s_d^2 = \sum_1^n (y_i - \hat{R}x_i)^2/(n - 1) = \sum_1^n [(y_i - \bar{y}) - \hat{R}(x_i - \bar{x})]^2/(n - 1)$$

$$= s_y^2 + \hat{R}^2 s_x^2 - 2\hat{R}s_{xy}$$

is an estimator for s_d^2. Hence, $V(\hat{R})$ may be approximately estimated from $v(\hat{R}) = (1 - f)s_d^2/n\bar{x}^2$.

Variance of the combined ratio estimator

The mean of $d_{gi} = y_{gi} - Rx_{gi}$ for the n_g units of the sample is $\bar{d}_g = \bar{y}_g - R\bar{x}_g$, and its mean for the N_g units of the gth stratum is $\bar{D}_g = \bar{Y}_g - R\bar{X}_g$. The population mean of d_{gi} is $\bar{D} = 0$. The variance of d_{gi} for the gth stratum is $S_{dg}'^2 = \Sigma(d_{gi} - \bar{D}_g)^2/(N_g - 1) = S_{yg}^2 + R^2 S_{xg}^2 - 2RS_{xyg}$. Since $\bar{d}_{st} = \Sigma W_g \bar{d}_g = \hat{\bar{Y}}_{RC} - R\bar{X}_{RC}$, (9.21) can be expressed as

$$V(\hat{\bar{Y}}_{RC}) = \sum W_g^2 V(\bar{d}_g) = \sum W_g^2 \frac{(1 - f_g)}{n_g} S_{dg}'^2.$$

With the sample variance $s_{dg}'^2 = \Sigma(d_{gi} - \bar{d}_g)^2/(n_g - 1) = s_{yg}^2 + \hat{R}_c^2 s_{xg}^2 - 2\hat{R}_c s_{xyg}$, an estimator for the variance in (9.21) is given by

$$v(\hat{\bar{Y}}_{RC}) = \sum W_g^2 \frac{(1 - f_g)}{n_g} s_{dg}'^2.$$

CHAPTER 10

Regression Estimation

10.1 Introduction

Scholastic success of students can be expected to depend to some extent on their scores on aptitude and achievement tests. Agricultural yields increase with the acreage and soil fertility. Industrial production may depend on employee sizes. Energy production and consumption of countries usually increase with their Gross National Products (GNPs). Consumer purchases frequently decrease with an increase in prices. Employment rates and stock prices are very much influenced by interest and inflation rates and related economic variables. Household expenses are directly related to family size.

In all these illustrations, the latter type of variables (x) provide auxiliary, concomitant, or supplementary information on the major variable (y). As the ratio method, the regression procedure utilizes this additional information to estimate the population total and mean of y with increased precision. For further benefits, this approach can also be combined with stratification. P.S.R.S. Rao (1987) presents a summary of the ratio and regression methods of estimation.

10.2 The regression estimator

In the illustration of Section 9.1 on test scores, the math score (y) is positively correlated with the total score (x). If the supplementary variable (x) and the major variable (y) are positively or negatively correlated, $(y_i - \bar{Y})$ can be expected to be very close to a constant multiple of $(x_i - \bar{X})$ for each of the population units. Consequently, departure of the sample mean of n units from the population mean, $(\bar{y} - \bar{Y})$, can be expected to be a multiple of the corresponding difference for the supplementary variable, $(\bar{x} - \bar{X})$. As a result, the **regression estimator** for $\bar{Y}$

that can be considered is

$$\hat{\bar{Y}}_l = \bar{y} + \beta(\bar{X} - \bar{x}),\tag{10.1}$$

where $\beta = S_{xy}/S_x^2 = \rho S_y/S_x$ is the population regression coefficient. This estimator is unbiased for $\bar{Y}$ and its variance is given by

$$V(\hat{\bar{Y}}_l) = V(\bar{y}) + \beta^2 V(\bar{x}) - 2\beta \; \text{Cov}(\bar{y}, \bar{x})$$
$$= \frac{(1-f)}{n}(S_y^2 + \beta^2 S_x^2 - 2\beta S_{xy}).\tag{10.2}$$

Note that $(S_y^2 + \beta^2 S_x^2 - 2\beta S_{xy}) = (S_y^2 - \beta^2 S_x^2) = S_y^2(1 - \rho^2) = S_e^2$, which is the error or residual mean square.

Clearly, $V(\hat{\bar{Y}}_l)$ becomes smaller than $V(\bar{y})$ as ρ becomes positively or negatively large. It is also smaller than $V(\hat{\bar{Y}}_R)$ in (9.4) unless β coincides with R, that is, the regression of y goes through the origin as described in Section 9.7.

> **Example 10.1.** Heights and weights: For the regression estimator of the average weight with height as the supplementary variable, from Table 9.1, $S_e^2 = 187.12(1 - 0.62^2) = 115.19$. For a sample of five from the 15 units, $V(\hat{\bar{Y}}_l) = 10(115.19)/75 = 15.36$ and hence S.E.$(\hat{\bar{Y}}_l) = 3.92$, which is slightly smaller than the S.E. of 3.97 found in Example 9.2 for the ratio estimator.

10.3   Estimation from the sample

From a sample (x_i, y_i), i = 1, 2,..., n, the slope β can be estimated from $b = s_{xy}/s_x^2$. Now, the regression estimator for $\bar{Y}$ is given by

$$\hat{\bar{Y}}_l = \bar{y} + b(\bar{X} - \bar{x})\tag{10.3}$$

The sample regression coefficient is not unbiased for β and as a result the estimator in (10.3) is biased for $\bar{Y}$. The bias of b and hence that of (10.3) become negligible for large n. The variance of (10.3) now becomes approximately the same as (10.2), and it can be estimated from

$$v(\hat{\bar{Y}}_l) = \frac{(1-f)}{n}s_e^2,\tag{10.4}$$

where s_e^2 is the residual mean square, which can be expressed as $(s_y^2 + b^2 s_x^2 - 2bs_{xy}) = (s_y^2 - b^2 s_x^2) = s_y^2 (1 - r^2)$. Note that $r = s_{xy}/s_x s_y$ is the sample correlation coefficient.

Example 10.2. Heights and weights: For the five sample units (4, 6, 9, 12, 15) considered in Examples 9.1 and 9.2, $\bar{x} = 66$, $\bar{y} = 158.6$, $s_y^2 = 314.8$, $s_x^2 = 24$, and $s_{xy} = 84.5$. Hence $b = 84.5/24 = 3.52$ and $\hat{Y}_l = 158.6 + 3.52(67.73 - 66) = 164.69$. Since $r = 84.5/(4.9 \times 17.74) = 0.97$, $v(\hat{Y}_l) = (2/15)(314.8)(1 - 0.97^2) = 2.48$, and hence S.E.$(\hat{Y}_l) = 1.57$.

10.4   Classical linear regression

As in the case of the ratio method, the linear regression procedure provides a motivation for the estimator in (10.3). In this approach, the principal characteristic y and the supplementary variable x are known as the dependent and independent or fixed variables. The **linear regression model** is

$$y_i = \alpha + \beta x_i + \varepsilon_i, \quad i = 1, 2,...,n. \tag{10.5}$$

The mean of y_i at x_i is given by $E(y_i | x_i) = \alpha + \beta x_i$, where α is the intercept and β is the slope of this regression line.

When $V(\varepsilon_i | x_i) = \sigma^2$ and ε_i and ε_j are uncorrelated, the LS estimators of the slope and intercept are obtained by minimizing $\Sigma(y_i - \alpha - \beta x_i)^2$ with respect to α and β. From this procedure, the estimators of these coefficients are given by

$$b = s_{xy}/s_{x^2} \quad \text{and} \quad a = \bar{y} - b\bar{x}. \tag{10.6}$$

These estimators are unbiased. The variance of b is given by $V(b) = \sigma^2/\Sigma(x_i - \bar{x})^2$. Noting that $\text{Cov}(b, \bar{y}) = 0$, $V(a) = [\sigma^2/n + \bar{x}^2\sigma^2/\Sigma(x_i - \bar{x})^2]$.

The residual SS, $\Sigma(y_i - a - bx_i)^2$ has $(n - 2)$ d.f. and it can be expressed as $\Sigma(y_i - \bar{y})^2 - b^2\Sigma(x_i - \bar{x})^2 = $ total SS $-$ regression SS. An unbiased estimator of σ^2 with $(n - 2)$ d.f. is given by the residual or error mean square, EMS $= \Sigma(y_i - a - bx_i)^2/(n - 2)$. Note that the residual mean square s_e^2 in (10.4) is not based on the model in (10.5) and it is obtained by dividing the residual sum of squares with $(n - 1)$.

The LS estimator for $E(y_i | x_i)$ and the predicted value of y_i at x_i are obtained from $(a + bx_i) = \bar{y} + b(x_i - \bar{x})$, which is the best linear unbiased estimator (**BLUE**). Predicting the N values y_i of a finite population, its mean is obtained from $\hat{Y}_l = \bar{y} + b(\bar{X} - \bar{x})$, which is the same as (10.3).

For the regression of the systolic pressures (y_i) with weight as the supplementary variable (x_i), from Table 9.1, $b = (0.92)(12.6)/13.68 = 0.8475$ and $a = 141.33 - 0.8475(155.53) = 9.52$. The mean $E(y_i \mid x_i)$ is estimated from $9.52 + 0.8475x_i$. The same equation provides the predicted values of y_i for specified x_i.

In general, if the regression of y_i on x_i is of the form in (10.5) and $V(\varepsilon_i \mid x_i) = \sigma^2 x_i^h = \sigma^2/W_i$, the WLS estimators of α and β are obtained by fitting the regression of $W_i^{1/2} y_i$ on $\alpha W_i^{1/2} + \beta W_i^{1/2} x_i$, that is, by minimizing $\Sigma W_i(y_i - \alpha - \beta x_i)^2$. Let $\bar{x}_w = \Sigma W_i x_i / \Sigma W_i$ and $\bar{y}_w = \Sigma W_i y_i / \Sigma W_i$ denote the weighted means. The estimators of the slope and intercept are given by $b = \Sigma W_i(x_i - \bar{x}_w)(y_i - \bar{y}_w)/\Sigma W_i(x_i - \bar{x}_w)^2$ and $a = \bar{y}_w - b\bar{x}_w$. These estimators are unbiased and their variances are given by $V(b) = \sigma^2/\Sigma W_i(x_i - \bar{x}_w)^2$ and $V(a) = \sigma^2[1/\Sigma W_i + \bar{x}_w^2/\Sigma W_i(x_i - \bar{x}_w)^2]$. An unbiased estimator of σ^2 is given by $\hat{\sigma}^2 = \Sigma W_i((y_i - a - bx_i)^2/(n - 2))$, which has $(n - 2)$ d.f. Both $b/\text{S.E.}(b)$ and $a/\text{S.E.}(a)$ follow Student's t-distribution with $(n - 2)$ d.f., and they can be used for testing $\beta = 0$ and $\alpha = 0$, respectively. The significance of the regression, that is, $\beta = 0$, can also be tested from $F = \text{regression MS/residual MS} = b^2\Sigma W_i(x_i - \bar{x}_w)^2/\hat{\sigma}^2$, which follows the F-distribution with 1 and $(n - 2)$ d.f.

The population mean $\bar{Y}$ can be estimated from $\hat{\bar{Y}}_w = \bar{y}_w + b(\bar{X} - \bar{x}_w)$. Its variance is given by $V(\hat{\bar{Y}}_w) = \sigma^2[1/\Sigma W_i + (\bar{X} - \bar{x}_w)^2/\Sigma W_i(x_i - \bar{x}_w)^2]$.

A graph of y_i plotted against x_i and examination of the variance of y_i at different values of x_i will be helpful in deciding on the suitable weighted ratio or regression type of estimator.

For the model in (10.5), Royall (1970) and Cochran (1977, pp. 199–200) show that the model-based optimum linear estimator takes the same form as $\hat{\bar{Y}}_l$ in (10.3). Royall (1976, 1986) and Valliant (1987) consider the model-based approach for the variance estimation for two-stage cluster sampling. Isaki and Fuller (1982), Särndal et al. (1992), Casady and Valliant (1993), and Tam (1986, 1995), among others consider the model-based approach for the estimation with sample survey data. Shah et al. (1977) describe the inference on the regression models from the data collected from surveys.

To estimate the population total Y or the mean $\bar{Y}$, the LS and the related model-based approaches result in linear combinations of the sample observations y_i, in general of the form $\Sigma l_i y_i$. Starting more than 45 years ago, Godambe (1955, 1966) made substantial theoretical contributions for finding optimum estimators utilizing all the available information on the population units that can be obtained, for example, from the auxiliary variables, the method of sampling, and the order of appearance of the population units in the sample.

10.5 Difference between regression estimators

Chapters 3 and 9 examined the difference between the sample means and the ratio estimators for two population means. The corresponding differences of the regression estimators can also be examined.

Common supplementary variable

Let $d_{1i} = y_{1i} + b_1(\bar{X} - x_i)$ and $d_{2i} = y_{2i} + b_2(\bar{X} - x_i)$. The sample means obtained from these expressions, $\bar{d}_1$ and $\bar{d}_2$, are the same as the regression estimators $\hat{\bar{Y}}_{l1} = \bar{y}_1 + b_1(\bar{X} - \bar{x})$ and $\hat{\bar{Y}}_{l2} = \bar{y}_2 + b_2(\bar{X} - \bar{x})$. The difference $(\hat{\bar{Y}}_{l1} - \hat{\bar{Y}}_{l2})$ is approximately unbiased for $(\bar{Y}_1 - \bar{Y}_2)$ and its variance can be estimated from

$$v(\hat{\bar{Y}}_{l1} - \hat{\bar{Y}}_{l2}) = \frac{1-f}{n}\Sigma[(d_{1i} - d_{2i}) - (\bar{d}_1 - \bar{d}_2)]^2(n-1)$$

$$= \frac{(1-f)}{n}\Sigma[(y_{1i} - \bar{y}_1) - (y_{2i} - \bar{y}_2)$$

$$-(b_1 - b_2)(x_i - \bar{x})]^2/(n-1). \tag{10.7}$$

Example 10.3. Systolic and diastolic pressures: With the data in Table 9.1, from the sample units (4, 6, 9, 12, 15), for the regression of systolic pressures on weight, $b_1 = 0.7195$ and hence $\hat{\bar{Y}}_{l1} = 150.21$. Similarly, for the diastolic pressures, $b_2 = 0.5051$ and $\hat{\bar{Y}}_{l2} = 94.21$. Thus, $(\hat{\bar{Y}}_{l1} - \hat{\bar{Y}}_{l2}) = 56$. From (10.7), the sample variance and S.E. of this estimate are 0.4702 and 0.6857.

Different supplementary variables

For this case, let $d_{1i} = y_{1i} + b_1(\bar{X}_1 - x_{1i})$ and $d_{2i} = y_{2i} + b_2(\bar{X}_2 - x_{2i})$. The corresponding sample means, $\bar{d}_1$ and $\bar{d}_2$ are the same as the regression estimators $\hat{\bar{Y}}_{l1} = \bar{y}_1 + b_1(\bar{X}_1 - \bar{x}_1)$ and $\hat{\bar{Y}}_{l2} = \bar{y}_2 + b_2(\bar{X}_2 - \bar{x}_2)$. The difference between these two estimators, $(\hat{\bar{Y}}_{l1} - \hat{\bar{Y}}_{l2})$, is approximately unbiased for $(\bar{Y}_1 - \bar{Y}_2)$, and its variance can be estimated from

$$v(\hat{\bar{Y}}_{l1} - \hat{\bar{Y}}_{l2}) = [(1-f)/n]\Sigma(e_{1i} - e_{2i})^2/(n-1), \tag{10.8}$$

where $e_{1i} = (y_{1i} - \bar{y}_1) - b_1(x_{1i} - \bar{x}_1)$ and $e_{2i} = (y_{2i} - \bar{y}_2) - b_2(x_{2i} - \bar{x}_2)$.

Example 10.4. Systolic pressure reduction with exercise: For the systolic pressures, as seen in Example 10.3, $\hat{Y}_l = 150.21$. For the observations after the fitness program, $s_{xy} = 210.15$ and $b = 210.15/279.8 = 0.7511$. If the average weight after the program is $\bar{x}_2 = 150$, $\hat{Y}_{l2} = 144.2 + 0.7511(150 - 151.4) = 143.15$. Thus, the difference of the regression estimators before and after the program is $150.21 - 143.15 = 7.06$. From (10.8), the estimate of the variance is 0.0459, and hence it has an S.E. of 0.2143.

10.6   Regression estimation vs. stratification

With the observations in Table T8 in the Appendix, it was found in Section 9.8 that for estimating the average wheat production in 1997, $V(\hat{Y}_{st}) = 0.84$ for samples of size four from each of the two strata. For the ratio estimation with the 1995 observations providing the supplementary information, $V(\hat{Y}_R) = 0.213$ for a sample of eight countries.

The observed data and the regression line $\hat{y}_i = 0.6264 + 1.0157x_i$ are presented in Figure 10.1. For this data, $S_e^2 = 2.6$ and for a sample of eight countries $V(\hat{Y}_l) = (12/160)2.6 = 0.195$ and S.E.$(\hat{Y}_l) = 0.44$.

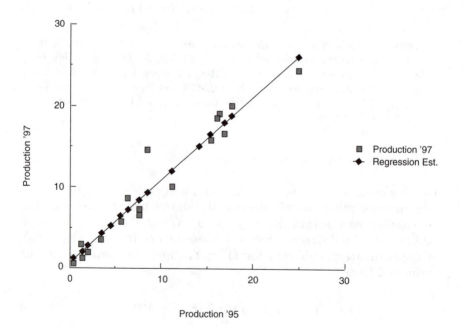

Figure 10.1.   Regression of 1997 wheat production on 1995 production.

10.7 Stratification and regression estimator

Similar to the procedures in Section 9.9 for the ratio estimation, one can obtain the separate or combined regression estimators for the population total and mean.

Separate estimator

For the gth stratum, the regression estimate for $\bar{Y}_g$ is $\hat{\bar{Y}}_{gl} = \bar{y}_g + b_g(\bar{X}_g - \bar{x}_g)$, where b_g is the regression coefficient for the gth stratum. The separate estimator for the population mean is

$$\hat{\bar{Y}}_s = (N_1 \hat{\bar{Y}}_{1l} + N_2 \hat{\bar{Y}}_{2l} + \cdots + N_G \hat{\bar{Y}}_{Gl})/N$$

$$= W_1 \hat{\bar{Y}}_{1l} + W_2 \hat{\bar{Y}}_{2l} + \cdots + W_G \hat{\bar{Y}}_{Gl} = \Sigma W_g \hat{\bar{Y}}_{gl}. \qquad (10.9)$$

Note that the numerator of the first expression is the regression estimator for the total.

For large n_g, the bias of this estimator will be negligible and its variance is approximately given by

$$V(\hat{\bar{Y}}_s) = \sum_{g=1}^{G} a_g S_{eg}^2, \qquad (10.10)$$

where $a_g = W_g^2 (1 - f_g)/n_g$ and $S_{eg}^2 = S_{yg}^2 (1 - \rho_g^2)$. For the estimate of this variance, replace S_{eg}^2 by $s_{eg}^2 = s_{yg}^2 - b_g^2 s_{xg}^2 = s_{yg}^2 (1 - r_g^2)$.

For Neyman allocation of a sample of size n, the sizes n_g in the strata should be chosen proportional to $N_g S_{eg}$. If ρ_g are close to each other, n_g can be chosen proportional to $N_g S_{yg}$.

Example 10.5. Wheat production: For the first stratum considered in Section 9.7, the regression line is $\hat{y}_i = 0.4985 + 0.9591 x_i$, with the S.E. of 0.46 and 0.099 for the intercept and slope. The corresponding t-statistics with $11 - 2 = 9$ d.f. are $t = 0.4985/0.46 = 1.1$ and $t = 0.9591/0.099 = 9.8$. From these figures, the intercept is not significantly different from zero at a 0.05 level of significance, but the slope is significantly different from zero at a level of significance of 0.005 or smaller.

For the second stratum, the regression line is $\hat{y}_i = 4.92 + 0.771 x_i$, with the S.E. of 2.61 and 0.16 for the intercept and the slope. Judging from $t = 4.92/2.61 = 1.88$ and $t = 0.771/0.16 = 4.8$, each following the t-distribution with $9 - 2 = 7$ d.f., the intercept and slope are significantly different from zero at the levels of significance of 0.10 and 0.001, respectively.

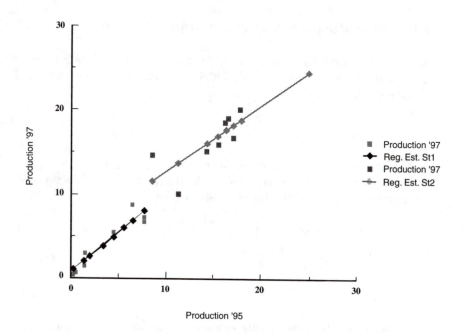

Figure 10.2. Regression of the wheat production for the two strata.

These regression lines are presented in Figure 10.2. From the data in Table T8 in the Appendix or the summary figures in Table 9.2, $s_{e1}^2 = 0.6386$ and $s_{e2}^2 = 3.5961$. Now, for samples of size four from the two strata, $v(\hat{\bar{Y}}_S) = (11/20)^2(7/44)(0.6386) + (9/20)^2 (5/36)(3.5961) = 0.1319$, which is much smaller than the variance of 0.20 for the separate ratio estimator.

Combined regression estimator

If it is known that the slopes β_g in the strata do not differ much, one can consider the combined estimator:

$$\hat{\bar{Y}}_C = \Sigma W_g[\bar{Y}_g + \beta(\bar{X}_g - \bar{x}_g)] = \hat{\bar{Y}}_{st} + \beta(\bar{X} - \hat{\bar{X}}_{st}), \qquad (10.11)$$

where β is the common slope. The variance of this unbiased estimator is

$$V(\hat{\bar{Y}}_C) = \sum_{g=1}^{G} a_g(S_{yg}^2 + \beta^2 S_{xg}^2 - 2\beta S_{xyg}). \qquad (10.12)$$

This variance is minimized when $\beta = \Sigma t_g \beta_g / \Sigma t_g$, where $t_g = a_g S_{xg}^2$. With this optimum slope, the variance in (10.12) becomes

$$V\left(\hat{\bar{Y}}_C\right) = \sum a_g (S_{yg}^2 - \beta^2 S_{xg}^2). \qquad (10.13)$$

From the samples in the strata, an estimator for β is $b = \Sigma t_g b_g / \Sigma t_g$. The estimator in (10.11) with β replaced by b is biased. For large samples, the bias will be small and the variance of $\hat{\bar{Y}}_C$ will be the same as (10.13).

With proportional allocation of the sample, b takes the form $\Sigma w_g s_{xyg} / \Sigma w_g s_{xg}^2$. The numerator and denominator of this expression are the weighted sum of cross-products and sum of squares, respectively. An estimate of the variance in (10.13) is

$$v(\hat{\bar{Y}}_C) = \sum a_g (s_{yg}^2 - b^2 s_{xg}^2). \qquad (10.14)$$

For the two strata considered above, $a_1 = 0.048$, $a_2 \doteq 0.028$, $t_1 = 0.137$, $t_2 = 0.128$. From these figures, $b = 0.8713$ and $v(\bar{Y}_C) = 0.104$. This variance is smaller than the variance of 0.1319 for the separate regression estimator.

Preliminary analysis and tests of hypotheses

To examine whether the separate or combined estimator is appropriate, one can plot the observations of the entire population and of the strata, examine the variance of y_i at different values of x_i, and fit the suitable regressions as described at the end of Section 10.4. As a next step, one should test the hypotheses that the slopes and intercepts are significantly different from zero. As a final step, one should test the hypothesis that the regressions for the strata are not the same; that is, their slopes as well as intercepts are different.

The final step for the regressions of the wheat productions is next examined. The residual SS for the separate regressions is $SS_{sep} = 7.1 + 29.4 = 36.5$ with $9 + 7 = 16$ d.f. and the corresponding mean square is $MS_{sep} = 36.5/16 = 2.28$. For the regression with all the 20 observations, the residual SS is 49.49 with 18 d.f. The difference in the two SS is $SS_{diff} = 49.49 - 36.5 = 12.99$ with $18 - 16 = 2$ d.f. and the corresponding mean square $MS_{diff} = 12.99/2 = 6.45$. The ratio

$F = MS_{\text{diff}}/MS_{\text{sep}}$ follows the F-distribution with 2 and 16 d.f. For this illustration, $F = 6.45/2.28 = 2.83$. From the tables of this distribution, $F = 7.51$ for 2 and 16 d.f. when the significance level is 0.05. Thus, there is no significant difference between the regressions for the two strata, which suggests the combined regression estimator.

10.8   Multiple regression estimator

The regression method of estimation can include more than one supplementary variable. For example, one can estimate the average wheat production in 1997 utilizing the supplementary data from 1995 and 1990.

With two supplementary variables, x_1 and x_2, the population observations can be represented by (y_k, x_{1k}, x_{2k}), $k = 1, 2,..., N$. Denote the means of these three variables by $(\bar{Y}, \bar{X}_1, \bar{X}_2)$, variances by (S_{00}, S_{11}, S_{22}), and their covariances by (S_{01}, S_{02}, S_{12}). For a sample of size n, denote the means by $(\bar{y}, \bar{x}_1, \bar{x}_2)$, variances by (s_{00}, s_{11}, s_{22}), and their covariances by (s_{01}, s_{02}, s_{12}).

The multiple regression estimator for the population mean of $\bar{Y}$ is

$$\hat{\bar{Y}}_{MI} = \bar{y} + \beta_1(\bar{X}_1 - \bar{x}_1) + \beta_2(\bar{X}_2 - \bar{x}_2), \qquad (10.15)$$

where β_1 and β_2 are the population *regression coefficients* or *slopes*. This estimator is unbiased and its variance is minimized when $\beta_1 = (S_{22}S_{01} - S_{12}S_{02})/(S_{11}S_{22} - S_{12}^2)$ and $\beta_2 = (S_{22}S_{02} - S_{12}S_{01})/(S_{11}S_{22} - S_{12}^2)$. With these optimum values, the variance of (10.14) becomes

$$V(\hat{\bar{Y}}_{MI}) = \frac{(1-f)}{n}S_e^2, \qquad (10.16)$$

where $S_e^2 = \Sigma[(y_i - \bar{Y}) - \beta_1(x_{1i} - \bar{X}_1) - \beta_2(x_{2i} - \bar{X}_2)]^2/(N-1)$ is the residual mean square (MS).

From the sample, the estimators of β_1 and β_2 are given by $b_1 = (s_{22}s_{01} - s_{12}s_{02})/(s_{11}s_{22} - s_{12}^2)$ and $b_2 = (s_{22}s_{02} - s_{12}s_{01})/(s_{11}s_{22} - s_{12}^2)$. Estimating the slopes, (10.15) becomes

$$\hat{\bar{Y}}_{M1} = \bar{y} + b_1(\bar{X}_1 - \bar{x}_1) + b_2(\bar{X}_2 - \bar{x}_2). \qquad (10.17)$$

This estimator, however, is not unbiased for the population mean. For large samples, its bias becomes negligible and its variance is approximately given by (10.16). An estimator of this variance is

$$v(\hat{\bar{Y}}_{Ml}) = \frac{(1-f)}{n} s_e^2, \tag{10.18}$$

where $s_e^2 = \Sigma[(y_i - \bar{y}) - b_1(x_{1i} - \bar{x}_1) - b_2(x_{2i} - \bar{x}_2)]^2/(n-1)$ is the sample residual MS.

The model for the classical multiple linear regression is

$$y_i = \alpha + \beta_1 x_{1i} + \beta_2 x_{2i} + \varepsilon_i, \tag{10.19}$$

where β_1 and β_2 are the *slope* coefficients and α is the *intercept*. For given values x_{1i} and x_{2i} of the *independent variables*, the expectation of the residual ε_i is assumed to be zero; that is, $E(\varepsilon_i | x_{1i}, x_{2i}) = 0$, and its variance is assumed to be σ^2, the same at different values of the independent variables. For some applications, the last assumption for the variance may not be valid and it is suitably modified. Further, ε_i and ε_j are assumed to be uncorrelated. The expectation of the *dependent variable* is $E(y_i | x_{1i}, x_{2i}) = \alpha + \beta_1 x_{1i} + \beta_2 x_{2i}$, which is the equation to a plane.

From the least squares principle, the slopes are estimated from b_1 and b_2, and the intercept from $a = \bar{y} - b_1 \bar{x}_1 - b_2 \bar{x}_2$. The estimator for the plane is given by $a + b_1 x_{1i} + b_2 x_{2i}$. The same expression is used for predicting y_i for specified (x_{1i}, x_{2i}). The residual mean square, $\Sigma[(y_i - \bar{y}) - b_1(x_{1i} - \bar{x}_1) - b_2(x_{2i} - \bar{x}_2)]^2/(n-3)$, is an unbiased estimator for σ^2.

The estimator in (10.15) is also obtained by considering the finite population as a sample from an infinite superpopulation, and predicting y_i with the above procedure. This approach provides another motivation for (10.15).

The following example examines whether it is advantageous to include both supplementary variables for estimation of the mean of the main characteristic.

Example 10.6. Wheat production: For the wheat production in 1997 (y) with the supplementary information from 1995 (x_1) and 1990 (x_2), from the data in Table T8 in the Appendix and Table 9.2, $a = 0.61$,

$b_1 = 0.869$, and $b_2 = 0.129$, with S.E. 0.60, 0.12, and 0.09, respectively. Each of the t-statistics $0.61/0.60 = 1.02$, $0.869/0.12 = 7.55$, and $0.129/0.09 = 1.43$ have $20 - 3 = 17$ d.f. From these results, only the slope coefficient for 1995 (x_1) is significantly different from zero.

From the above data, $S_e^2 = 2.3253$ and for a sample of eight countries, $V(\hat{Y}_{Ml}) = 12(2.3253)/160 = 0.1744$ and S.E.$(\hat{Y}_{Ml}) = 0.42$. This variance and S.E. are only a little different from the S.E. of 0.44 found in Section 10.6 for the regression estimator with the supplementary information from 1995. The above results from the classical tests of hypotheses also lead to the same conclusion.

As in Section 10.7, the multiple regression estimator can also be considered along with stratification.

10.9   Double sampling regression estimator

If $\bar{X}$ is not known, as described in Section 9.11 for the ratio estimator, the double-sampling regression estimator for $\bar{Y}$ is obtained from

$$\hat{\bar{Y}}_{ld} = \bar{y} + b(\bar{x}_1 - \bar{x}), \qquad (10.20)$$

where $\bar{x}_1$ is the mean of the first sample of n_1 units and $(\bar{x}, \bar{y})$ as well as b are obtained from the second sample of n units selected from the first sample. Following Appendix A3, for large n, the bias of this estimator becomes negligible and its variance becomes

$$V(\hat{\bar{Y}}_{ld}) = (N - n_1)S_y^2/Nn_1 + (n_1 - n)S_y^2(1 - \rho^2)/n_1 n. \qquad (10.21)$$

For the case of selecting the second sample independent of the first, P.S.R.S. Rao (1972) replaces $\bar{x}_1$ in (10.20) by $\bar{x}_a = a\bar{x}_1 + (1 - a)\bar{x}$ or the mean of the distinct units $\bar{x}_v$, as described in Section 9.11 for the ratio estimator.

The regression estimator for the rotation surveys take the form of (10.20) or its modifications. Through the model in (10.5), Dorfman (1994) derives a model-based variance estimator for (10.20). P.S.R.S. Rao (1998b) presents a brief summary on the double sampling regression estimators.

10.10   Generalized regression and calibration estimators

When the n sample units are selected with probabilities ϕ_i, to estimate the population total Y, Cassel et al. (1976) suggest the **generalized regression estimator** (GREG)

$$\hat{Y}_{greg} = \Sigma(y_i/\phi_i) + b[X - \Sigma(x_i/\phi_i)], \qquad (10.22)$$

where $b = \Sigma(x_i y_i/\phi_i)/\Sigma(x_i^2/\phi_i)$.

This estimator can be expressed as $\Sigma d_i y_i$. To relate ϕ_i with a supplementary characteristic, Deville and Sarndal (1992) consider the **calibration estimator** $\hat{Y}_c = \Sigma w_i y_i$. The weights w_i are found by minimizing $\Sigma[(w_i - d_i)^2/d_i]$ with the constraint $\Sigma w_i x_i = X$. Wright (1983) combines the features of both the GREG and calibration methods to estimate Y.

Exercises

10.1.   *Project.* Consider the 20 samples of size three of Exercise 2.10 for regression estimation of the mean of the math scores with the verbal scores as the supplementary variable. Find the expectation and bias of (a) the slope b and (b) the regression estimator $\hat{Y}_l$ for the mean of the math scores. (c) Compare the exact variance and MSE of $\hat{Y}_l$ with the approximate variance in (10.2). (d) Find the expectation of the approximate variance in (10.4) and its bias for estimating the variance and MSE of $\hat{Y}_l$.

10.2.   From the five sample units (4, 6, 9, 12, 15) selected from Table 9.1, consider the regression estimation for the mean of the diastolic pressures. Find (a) the estimate, (b) its S.E. and (c) the 95% confidence limits for the mean.

10.3.   With the sample units (4, 6, 9, 12, 15), as in Example 10.3, find (a) the difference of the regression estimates of the means of the systolic and diastolic pressures utilizing weight as the supplementary variable, (b) the S.E. for the difference of the estimates, and (c) the 95% confidence limits for the difference of the means.

10.4.   For the sample units (4, 6, 9, 12, 15), consider the diastolic pressures (80, 92, 95, 80, 93) after the fitness program.

(a) Estimate the change in the diastolic pressures after the fitness program through the regression method and (b) find the 95% confidence limits for the change.

10.5. *Project.* From the data in Table T4 in the Appendix for the largest enrollment, for 1990 and 1995, for a sample of size four, (a) find the variance of the regression estimators for the average of the public enrollment with total enrollment as the supplementary variable, and (b) compare these variances with those found in Exercise 9.11 for ratio estimators.

10.6. Find the variance of the difference of the regression estimators in Exercise 10.5 and compare it with the variance of the difference of the sample means.

10.7. From the data in Table T5 in the Appendix for the smallest enrollment, for a sample of size four, (a) find the variance of the difference of the regression estimators for 1995 for public and private enrollment with total enrollment as the supplementary variable, and (b) compare it with the variance of the difference of the sample means.

10.8. Divide the data in Table 9.1 into two strata with the ranges 63 to 66 and 68 to 72 inches for the heights, and consider the corresponding heights, weights, and blood pressures. (a) Fit the regressions of weight on height in the two strata and perform the test of the hypothesis for their difference. (b) Irrespective of the inference from the test in (a), find the separate and combined regression estimators for the average weight, and compare their standard errors. (c) Examine whether or not the conclusion in (b) agrees with the inference in (a).

10.9. *Project.* As in Exercise 9.14, compute the separate and combined regression estimates in (10.9) and (10.11). (a) From the averages and variances of the estimates, find the biases and MSEs for the two procedures. (b) Compare the exact MSEs in (a) with the approximate expressions in (10.10) and (10.12).

10.10. From the data of the 25 countries in Table T8 in the Appendix for wheat production, fit the regressions of the production for 1997 (y) on 1995 (x_1), on 1990 (x_2), and on both x_1 and x_2. To estimate the total production in 1997 for the 25 countries with a sample of 10 countries, find the standard errors of the regression estimators with x_1, x_2 and (x_1,x_2) providing the supplementary information.

10.11. Based on the wheat production in 1997, the 25 countries in Table T8 in the Appendix can be divided into three strata consisting of the first 11, the next 9, and the remaining 5 countries. To estimate the total production in 1997 (y) with 1995 (x) as the supplementary variable, find the S.E. of the separate regression estimator for samples of 4, 4, and 2 countries from the three strata.

CHAPTER 11

Nonresponse and Remedies

11.1 Introduction

In almost every survey, some of the persons, households, and other types of units selected into the sample are not contacted. Persons away from home on business or vacation, wrong addresses and telephone numbers, households without telephones or with unlisted numbers, inability of the interviewers to reach households in remote places, and similar reasons contribute to the noncontacts. Even if they are contacted, some of the units may not respond to one or more characteristics of the survey. As noted in Section 3.7, estimators obtained from the responding units alone are biased, and increasing the sample size may reduce their variances but not the biases.

Public polls are also frequently affected by the **noncontacts** and nonresponse. In the Truman–Dewey presidential contest, although Truman emerged as the winner from the final count of the ballots, *The Literary Digest*, a newspaper in Chicago, initially declared Dewey the winner. A badly conducted poll and a large amount of nonresponse were blamed for this "fiasco." Mosteller et al. (1949) analyzed the effects of nonresponse in this survey.

Cochran (1977, Chap. 12) presents some of the reasons for nonresponse and the procedures to counteract its effects. A review of progress in sample surveys including efforts to reduce the bias arising from nonresponse was provided by Cochran (1983). Godambe and Thompson (1986) describe theoretically optimum procedures of estimation in the presence of nonresponse. Little and Rubin (1987) present the general methods available for estimating missing observations in statistical designs of experiments, regression analysis, and other types of statistical procedures.

Surveys are also affected by errors of observation and measurement. Nonresponse together with these types of errors are known as **nonsampling errors**. Mahalanobis (1946) recommends the procedure of **interpenetrating subsamples**, in which independent samples are

assigned to different interviewers. The above types of errors are
assessed from the variation among the estimates of the interviewers.
Lessler and Kalsbeek (1992) describe the effects of nonresponse and
errors of measurements. Biemer et al. (1991) examine the effects of
measurement errors.

Some of the exercises of previous chapters have examined simple
procedures of adjusting estimates for nonresponse. In the following
sections, the reasons for nonresponse and the procedures for reducing
its effects are presented in detail.

11.2   Effects of survey topics and interviewing methods

Some of the units selected into the sample may not respond to the
entire survey and some may provide answers only to selected items in
the questionnaire. These two cases are classified as **unit nonresponse**
and **item nonresponse**, respectively. A number of factors affect
response rates, and some of the main reasons are described below.

Subject matter

To a large extent, response rates depend on the interest of the sampled
units in the subject matter of the survey. Response rates of the public
tend to be high on matters related to educational reforms, local
improvements, political views, tax changes, and similar characteristics. Concerned citizens can be expected to respond readily to topics
of national interest. Low response rates are frequently observed in
surveys on personal characteristics, such as incomes and savings.

Types of interviewing

Surveys are conducted through the mail (post), e-mail, telephone, face-
to-face personal interviews, or a combination of these procedures.
Dillman (1978) describes surveys conducted through the mail and
telephone. Nonresponse in telephone surveys is described in the book
edited by Groves et al. (1988). Collins (1999) summarizes the response
rates of the telephone interviews in the U. K. Sirken and Casady (1982)
describe the nonresponse rates in the dual frame surveys in the U.S.
in which samples are selected from the lists and telephone numbers of

the population units. From the Canadian Labour Force Survey, Gover (1979) notes that response rates can differ with the interviewers.

In the **random digit-dialing** method, the telephone digits of people to be included in the sample are selected randomly; see Waksberg (1978), for example. This approach is practiced in some types of marketing surveys and for public polls, as an attempt to obtain responses from the persons with both listed and unlisted telephone numbers. Computer-aided telephone-interviewing (CATI) is being tested in both industrialized and developing nations; see Groves and Nicholls (1986), for example. Martin et al. (1993) describe Computer-aided personal interviewing (CAPI). Response rates have been found to be different for all these methods of conducting the surveys.

In some surveys, telephone reminders and follow-up letters are employed to increase response rates. In **callback** surveys, more than one attempt is made to contact and elicit responses from the sample units. Personal interviews are usually found to yield higher response rates than mail and telephone surveys.

Types of respondents

In some household surveys, responses to the survey questions are requested from **any adult** in the household. For some types of surveys, **specified** persons such as parents, heads of households, supervisors, and managers are asked to provide the required information. In some other types of surveys, the respondent is **randomly chosen** from all the persons eligible to provide answers to the items of the survey. Response rates are usually found to be relatively higher for the first type of respondent and lower for the last.

11.3 Response rates

A number of studies summarize the response rates of surveys. For example, Bailar and Lanphier (1978) found the response rates for 36 surveys conducted by the government and private organizations to be low; for three of these surveys, the rates were only 25, 46, and 50% in spite of several attempts to obtain the responses. From analyzing a large number of surveys, P.S.R.S. Rao (1983c) found that response rates for the *any*, *specified*, and *random adults* at the initial attempt were (69.4, 51.7, 33.1)% and (94.9, 89.1, 83.9)% after three or more attempts.

11.4   Bias and MSE

To examine the bias due to nonresponse, the population may be considered to consist of two strata or groups, with **respondents** and **nonrespondents**. Members of the first group provide answers to a question on the survey, if they were contacted. The second group does not provide answers even after repeated attempts.

In Section 3.7, the size of the responding group, their mean, and variance are denoted by N_1, $\bar{Y}_1$, and S_1^2, respectively. Similarly, $N_2 = N - N_1$, $\bar{Y}_2$, and S_2^2 denote the corresponding figures for the nonrespondents. The population mean now can be expressed as $\bar{Y} = W_1 \bar{Y}_1 + W_2 \bar{Y}_2$, where $W_1 = N_1/N$ and $W_2 = N_2/N$.

The number of respondents and nonrespondents in a simple random sample of n units from the population are denoted by n_1 and $n_2 = n - n_1$. The sample mean $\bar{y}_1$ is unbiased for $\bar{Y}_1$. As shown in Section 3.7, its bias, variance, and MSE for estimating $\bar{Y}$ are given by $B(\bar{y}_1) = \bar{Y}_1 - \bar{Y} = W_2(\bar{Y}_1 - \bar{Y}_2)$ and $V(\bar{y}_1) = (1 - f_1)S_1^2/n_1$ and MSE $(\bar{y}_1) = V(\bar{y}_1) + B^2(\bar{y}_1)$, where $f_1 = n_1/N_1$.

An unbiased estimator for the total $Y_1 = N_1\bar{Y}_1$ of the respondents is $\hat{Y}_1 = N_1\bar{y}_1$. Since N_1 is not known, an unbiased estimator can be obtained by replacing it with $N(n_1/n)$.

The sample ratio $\hat{R}_1 = \bar{y}_1/\bar{x}_1$ is approximately unbiased for $R_1 = \bar{Y}_1/\bar{X}_1$ of the respondents. Its bias in estimating $R = \bar{Y}/\bar{X}$ will be small only if $R_2 = \bar{Y}_2/\bar{X}_2$ for the nonresponse group does not differ much from R_1. Similarly, the expected value of the regression estimator $\bar{y}_1 + b_1(\bar{X} - \bar{x}_1)$ is approximately equal to $\bar{Y}_1 + \beta_1(\bar{X} - \bar{X}_1)$, where β_1 and b_1 are the population and sample slopes for the respondents. Hence, its bias will be negligible only if the respondents and nonrespondents do not differ much in the means of x and y and the slopes.

In some situations, the n_1 responses may be considered to be a random sample from the n sampled units. This will be the case, for example, if the interviewer elicits responses through intensive efforts from a random selection of n_1 of the n sampled units. For such a situation, $\bar{y}_1$ is unbiased for $\bar{Y}$ and its variance is given by $V(\bar{y}_1) = (N - n_1)S^2/Nn_1$. An unbiased estimator of this variance is obtained by replacing S^2 by the sample variance of the respondents, $s_1^2 = \Sigma_1^{n_1}(y_i - \bar{y}_1)^2/(n_1 - 1)$.

11.5   Estimating proportions

As in Chapter 4, let C and $P = C/N$ denote the number and proportion of units in the population having a specified attribute. Similarly, let C_1 and C_2 denote the numbers of units having the attribute in the

response and nonresponse groups. The corresponding proportions are
denoted by $P_1 = C_1/N_1$ and $P_2 = C_2/N_2$.

Among the n_1 respondents of the sample, c_1 will be observed to
have the attribute. The sample proportion $p_1 = c_1/n_1$ is unbiased for
P_1, and its bias for estimating the population proportion P is $W_2(P_1 -
P_2)$. The absolute value of this bias increases with the proportion of
the nonrespondents W_2 and the difference between P_1 and P_2.

An unbiased estimator of the variance of p_1 is $v(p_1) = (N_1 - n_1)p_1 \times
(1 - p_1)/N_1(n_1 - 1)$. Since N_1 is not known, it may be replaced by its
estimator Nn_1/n. With the normal approximation, $(1 - \alpha)\%$ confidence
limits for P_1 are obtained from

$$p_1 \pm Z\sqrt{v(p_1)}. \tag{11.1}$$

Example 11.1. Proportions: Consider a random sample of $n = 500$ units
from a large population in which only $n_1 = 300$ units respond. If 120 of
the respondents are observed to have the characteristic of interest, $p_1 =
120/300$ or 40%. The variance of this estimate is $0.4(0.6)/299 = 0.0008$,
and hence it has a S.E. of 0.0283. Approximate 95% confidence limits
for P_1 are given by $0.4 \pm 1.96(0.0283)$; that is, $(0.34, 0.46)$.

If one considers the n_1 responses to be a random sample from the n
units of the initial sample, p_1 is unbiased for P. In this case, the variance
of p_1 becomes $(N - n_1) PQ/(N - 1)n_1$, where $Q = 1 - P$. Now, the limits
$(0.34, 0.46)$ in the above example refer to P, the population proportion.

11.6   Subsampling the nonrespondents

Deming (1953) presented a model for studying the effectiveness of
callbacks. Through this model and the data from practical surveys,
P.S.R.S. Rao (1983c) found that for a variety of surveys on an average
three calls are required to obtain high response rates. Since callbacks
can be expensive, Hansen and Hurwitz (1946) suggest eliciting
responses from a subsample of the nonrespondents. For the case of
mail surveys at the initial attempt and personal interviews at the
second stage, they present the following estimator for the mean and
its variance. The procedure of subsampling, however, is applicable for
any method of conducting a survey. A summary of this approach
appears in P.S.R.S. Rao (1983b).

The procedure of Hansen and Hurwitz

As before, let N_1 and $N_2 = N - N_1$ denote the sizes of the response
and nonresponse strata. In a simple random sample of n units from

the population, n_1 responses and $n_2 = n - n_1$ nonresponses are obtained. A subsample of size $m = n_2/k$, with a predetermined value of k (>1), is drawn from the n_2 units and responses from all the m units are obtained.

Estimator and its variance

Let $\bar{y}_1$ and $\bar{y}_{2(m)}$ denote the means of the n_1 respondents at the first stage and the m subsampled units. An estimator for the population mean is given by

$$\bar{y}_H = w_1\bar{y}_1 + w_2\bar{y}_{2(m)},   \tag{11.2}$$

where $w_1 = n_1/n$ and $w_2 = n_2/n$. Note that for this estimator the mean of the m units at the second stage is inflated by w_2, and the mean $\bar{y}_2$ of the n_2 nonrespondents is not available.

For a given sample (s) consisting of n_1 respondents and n_2 nonrespondents $E(\bar{y}_H \mid s) = w_1\bar{y}_1 + w_2\bar{y}_2 = \bar{y}$, which is the mean of the n sampled units. Hence, $\bar{y}_H$ is unbiased for $\bar{Y}$, and from Appendix A11,

$$V(\bar{y}_H) = \frac{1-f}{n}S^2 + W_2\frac{k-1}{n}S_2^2.   \tag{11.3}$$

The second term is the increase in the variance due to subsampling the nonrespondents. This increase will be large if the size of the second stratum or its variance is large. It can be reduced by increasing the subsampling fraction $1/k$.

Optimum sample sizes

For the above procedure, the cost of sampling is considered to be of the form

$$E' = e_0n + e_1n_1 + e_2n_2,   \tag{11.4}$$

where e_0 is the initial cost for arranging for the survey. The cost for obtaining the required information from a respondent at the first and second stages are denoted by e_1 and e_2, respectively. From (11.4), the expected cost becomes

$$E = (e_0 + e_1 W_1 + e_2 W_2/k)n. \tag{11.5}$$

Minimizing the variance in (11.3) for given cost in (11.5), or the cost for given variance V, is the same as minimizing $(V + S^2/N)E$. From this minimization, the optimum value of k is given by

$$k = \left[\frac{e_2(S^2 - W_2 S_2^2)}{(e_0 + e_1 W_1)S_2^2} \right]^{1/2}. \tag{11.6}$$

Thus, the size of the subsample will be large if e_2 is large relative to $(e_0 + e_1 W_1)$. Since additional effort is needed to elicit responses from a subsampled nonrespondent, e_2 is usually larger than e_1. For a specified V, from (11.3) and (11.6), the optimum value for the size of the initial sample is obtained from

$$n = \frac{N[S^2 + (k - 1)W_2 S_2^2]}{NV + S^2}$$

$$= n_0 \left[1 + \frac{(k - 1)W_2 S_2^2}{S^2} \right], \tag{11.7}$$

where $n_0 = NS^2/(NV + S^2)$ is the sample size required when $W_2 = 0$. For fixed E, the optimum sample size from (11.5) and (11.6) is given by

$$n = \frac{kE}{k(e_0 + e_1 W_1) + e_2 W_2}. \tag{11.8}$$

For finding the optimum sizes of the samples at the two stages, it is enough to know the relative value S_2^2/S^2 of the variances of the nonrespondents and the population. The value of W_2, however, should be known.

Example 11.2. Sample sizes and costs for a specified precision: Consider a population of $N = 30{,}000$ units with the standard deviation $S = 648$ for a characteristic of interest. If there is no nonresponse, for a sample of $n_0 = 2000$ units from this population, the variance of the sample mean is $(N - n_0)S^2/Nn_0 = 196$.

One may require that when there is nonresponse, the variance in (11.3) should be the same, that is $V = 196$. For the costs, consider $e_0 = 0.5$, $e_1 = 1$, and $e_2 = 4$. Now, if $W_1 = 0.7$ and $S^2 = 1.8S_2^2$, from (11.6), $k = 2.23$ or $1/k = 0.45$; that is, 45% of the nonrespondents should be sampled at the second stage. From (11.7), the optimum size for the initial sample is $n = 2410$. From (11.5) this survey would cost on the average \$4189. Without nonresponse, it would cost $(e_0 + e_1)n_0$, that is, \$3000.

Since the subsampling fraction depends on W_2, Srinath (1971) suggests a modification, which was examined by P.S.R.S. Rao (1983a). Särndal and Swensson (1985) consider unequal probability sampling at both stages. The above procedure of subsampling nonrespondents can be extended to the case of stratification.

Variance estimation

For convenience, denote the mean $\bar{y}_{2(m)}$ and the variance of the m units $s_{2(m)}^2 = \Sigma^m(y_1 - \bar{y}_{2(m)})^2/(m-1)$ by $\bar{y}_m$ and s_m^2. This variance is unbiased for s_2^2 of the $n_2 = n - n_1$ nonrespondents and hence for S_2^2. The sample variance s^2 of the n units is unbiased for S^2, but it is not available due to the nonresponse. An unbiased estimator for S^2 and $V(\bar{y}_H)$ are obtained in Appendix A11 from the derivations of Cochran (1977, p. 333) and J.N.K. Rao (1973). For large N, this estimator becomes

$$v(\bar{y}_H) = [(n_1 - 1)s_1^2 + (n_2 - 1)ks_m^2 + n_1(\bar{y}_1 - \bar{y}_H)^2$$
$$+ n_2(\bar{y}_m - \bar{y}_H)^2]/n(n - 1). \qquad (11.9)$$

Ratio and regression estimators

As in the case of $\bar{Y}$, an unbiased estimator for the mean $\bar{X}$ of an auxiliary characteristic is obtained from

$$\bar{x}_H = w_1\bar{x}_1 + w_2\bar{x}_{2(m)}. \qquad (11.10)$$

In this expression, $\bar{x}_1$ and $\bar{x}_{2(m)}$ are the means of the n_1 respondents and the m subsampled units. Now, the ratio estimator for $\bar{Y}$ is given by

$$\bar{y}_{HR} = (\bar{y}_H/\bar{x}_H)\bar{X} = \hat{R}_H\bar{X}. \qquad (11.11)$$

For large n and m, following the approach in Appendix A11, the variance of $\bar{y}_{HR}$ is obtained from

$$V(\bar{y}_{HR}) = \frac{(1-f)}{n}S_d^2 + W_2\frac{(k-1)}{n}S_{d2}^2, \qquad (11.12)$$

where $S_d^2 = \Sigma_1^N(y_i - Rx_i)^2/(N-1)$ and $S_{d2}^2 = \Sigma_1^{N_2}(y_{2i} - Rx_{2i})^2/(N_2-1)$. The estimator in (11.11) and the variance in (11.12) were suggested by Cochran (1977, p. 374).

For the regression estimator of $\bar{Y}$, let

$$A = \sum_1^{n1}(x_{1i} - \bar{x}_1)(y_{1i} - \bar{y}_1) + (n_2 - 1)\sum_1^m(x_{2i} - \bar{x}_m)(y_{2i} - \bar{y}_m)/(m-1)$$

and

$$B = \sum_1^{n1}(x_{1i} - \bar{x}_1)^2 + (n_2 - 1)\sum_1^m(y_{2i} - \bar{y}_m)^2/(m-1). \qquad (11.13)$$

Note that the subscript m refers to the subsample. An estimator for the slope is given $b = A/B$. For $\bar{Y}$, one can now consider

$$\bar{y}_{Hl} = \bar{y}_H + b(\bar{X} - \bar{x}_H). \qquad (11.14)$$

For large n and m, the bias of this estimator is negligible and its variance approximately becomes

$$V(\bar{y}_{Hl}) = \frac{(1-f)}{n}S_e^2 + W_2\frac{(k-1)}{n}S_{e2}^2, \qquad (11.15)$$

where $S_e^2 = \Sigma_1^N[(y_i - \bar{Y}) - \beta(x_i - \bar{X})]^2/(N-1)$ and $S_{e2}^2 = \Sigma_1^{N_2}[(y_{2i} - \bar{Y}_2) - \beta(x_{2i} - \bar{X}_2)]^2/(N_2 - 1)$.

There may not be any nonresponse for auxiliary variables such as family size and years of education. In such cases, $\bar{x} = \Sigma_1^n x_i/n$ will be available and an alternative ratio estimator for $\bar{Y}$ is given by $(\bar{y}_H/\bar{x})\bar{X}$, and a similar regression estimator can be found. P.S.R.S. Rao (1987, 1990) examines the biases and MSEs of these types of estimators.

11.7   Estimating the missing observations

Procedures developed for estimating the missing observations in statistical designs of experiments and regression analysis can also be utilized for predicting the observations of the nonrespondents.

Least squares estimation

For statistical analysis, Yates (1933) recommended estimation of the missing observations through the least squares principle. For the regression model in (10.5), if x is observed on all the n units, but y is observed on only n_1 units, the n_2 missing values of y along with the coefficients (α, β) can be estimated by minimizing

$$\delta = \sum_1^{n_1}(y_i - \alpha - \beta x_i)^2 + \sum_1^{n_2}(y_i - \alpha - \beta x_i)^2. \qquad (11.16)$$

This optimization results in

$$\hat{\beta} = b_1 = \frac{s_{xy1}}{s_{x1}^2}, \qquad \hat{\alpha} = \bar{y}_1 - b_1\bar{x}_1$$

and

$$\hat{y}_i = \bar{y}_1 + b_1(x_i - \bar{x}_1), \qquad (11.17)$$

for $i = (1, ..., n_2)$. With these estimates,

$$\delta = \sum_1^{n_1}(y_i - \hat{\alpha} - \hat{\beta}x_i)^2 = \sum_1^{n_1}[(y_i - \bar{y}_1) - b_1(x_i - \bar{x}_1)]^2, \quad (11.18)$$

which is the same as the residual sum of squares computed from the n_1 completed observations on (x_i, y_i). Note that from (11.17), the average of the predictions for the n_2 missing observations becomes $\bar{y}_1 + b_1(\bar{x}_2 - \bar{x}_1)$.

For the regression through the origin, if $V(y_i | x_i)$ is proportional to x_i, following the model in (9.10),

$$\delta = \sum_1^{n_1}[(y_i - \beta x_i)^2/x_i] + \sum_1^{n_2}[(y_i - \beta x_i)^2/x_i] \qquad (11.19)$$

is minimized. The estimator of the slope β now is given by $b_1 = \bar{y}_1/\bar{x}_1$, and the observations of the n_2 nonrespondents are predicted from $b_1 x_i$. In this case, the average of the predictions for the n_2 missing observations becomes $(\bar{y}_1/\bar{x}_1)\bar{x}_2$.

Example 11.3. Survey on families: Table 11.1 presents data on five variables from a simple random sample of 16 from 2000 families, along with the number of responses, means, and standard deviations. For income, mid-values of the ranges (in 1000s) 15 to 20, 20 to 25,... are considered.

Table 11.1.   Survey of families.

No.	Family Size, x	Husband's Age, h	Wife's Age, w	Income Level, y	Television Time, t
1	3	35	32	62.5	15
2	4	45	45	59.5	12
3	5	45	—	48.5	15
4	3	28	26	22.5	12
5	5	48	—	72.5	18
6	5	50	—	49.5	12
7	2	26	26	47.5	10
8	3	28	25	37.5	12
9	4	36	32	52.5	—
10	2	27	23	22.5	—
11	4	44	—	62.5	—
12	2	29	29	32.5	—
13	4	46	38	—	14
14	3	32	28	—	12
15	5	44	—	—	15
16	6	45	35	—	14
Responses	16	16	11	12	12
Mean	3.75	38	30.82	47.50	13.42
S.D.	1.24	8.65	6.52	16.02	2.15

	Husband's Age			Wife's Age		
Source	d.f.	SS	F	d.f.	SS	F
Regression	1	1465.4	10.8	1	888.6	6.3
Residual	10	1358.6		6	849.3	
Total	11	2824		7	1737.9	

$$\hat{y}_i = 1.07 + 1.26 h_i \qquad\qquad \hat{y}_i = -6.0 + 1.62 w_i$$

S.E.	14.50	0.38	S.E.	19.7	0.65
t_{10}	0.07	3.28	t_6	−0.31	2.51
p-value	0.94	0.008	p-value	0.71	0.046

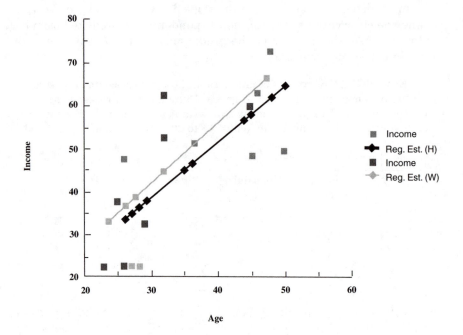

Figure 11.1. Regressions of income on husband's age and wife's age.

Regression of income on husband's age for the 12 pairs of completed observations and on wife's age for the eight pairs are presented in Figure 11.1, along with the summary figures needed for statistical analysis. For both the regressions, the slopes are significantly larger than zero but not the intercepts. If these regressions are still used from the first regression, prediction of the missing income at husband's age $h = 46$ is $1.07 + 1.26(46) = 59.03$.

For the 12 pairs of responses on both husband's age and income, $\bar{h}_1 = 36.75$ and $\bar{y}_1 = 47.5$. For the four nonresponses on income, the mean for husband's age is $\bar{h}_2 = 41.75$. Since $b_1 = 1.26$, prediction of the average income $\bar{y}_2$ for the four nonrespondents from the regression on husband's age is $47.5 + 1.26(41.75 - 36.75) = 53.8$.

One may now predict the missing observations on income from the regression on wife's age. At $w = 28$, prediction for the income is $-6 + 1.62(28) = 39.36$. Similarly, at wife's ages of 35 and 38, predictions for income are 50.7 and 55.56.

As described above, for the regression through the origin of income on husband's age, $b_1 = 47.5/36.75 = 1.29$. The prediction of income at $h = 46$ now is $1.29(46) = 59.34$, which is slightly larger than 59.03 for the regression with the intercept.

Similarly, for the regression through the origin of income on wife's age with the eight pairs of responses, $b_1 = 42.125/29.75 = 1.42$. Now, prediction for the income at $w = 28$ is $1.42(28) = 39.76$. At $w = 35$ and 38, the predictions are 49.7 and 53.96.

One should note that for this type of predicting the missing observations, it is assumed that the slopes as well as the intercepts for the responding and nonresponding group are the same. Further, in the above illustration, multiple regression of income on both husband's and wife's age can be attempted, but it will be based on only eight completed sets of observations.

Alternative predictions

In the procedure suggested by Buck (1960), the missing observations on each variable are predicted by fitting its regression on the remaining variables, using the completed observations on all the variables. For example, in the case of two variables x and y, the observations on y are predicted from the regression of y on x. Similarly, the missing observations on x are predicted from its regression on y. Schafer (1997) describes statistical analysis for the case of missing observations on more than one variable.

Hendricks (1949) made one of the earliest suggestions for predicting the percentage for an attribute from the trend of the percentages at the successive attempts in a mail survey. As an illustration, suppose the response rates for three successive attempts are $(60, 20, 10)\%$, with the corresponding percentages $(54, 47, 52)$ favoring an attribute in the survey. For the regression of $y_i = (54, 47, 52)$ on $x_i = (1, 2, 3)$, the slope and intercept coefficients are $b = -0.5$ and $a = 52$. Hence, the equation for predicting the percentage at the successive attempts is $y_i = 52 - 0.5x_i$. From this expression, at $x_i = 4$, $y_i = 52 - 0.5(4) = 50$. Now, an estimate for the population percentage favoring the attribute is $54(0.60) + 47(0.20) + 52(0.10) + 50(0.10) = 52$.

11.8 Ratio and regression estimation

Ratio estimation

If $\bar{X}$ is known, a ratio-type estimator for $\bar{Y}$ is $\bar{y}_R = (\bar{y}_1/\bar{x}_1)\bar{X} = \hat{R}_1\bar{X}$, where $(\bar{x}_1, \bar{y}_1)$ are the means of the n_1 respondents and $\hat{R}_1 = \bar{y}_1/\bar{x}_1$. If the respondents are considered to be a random sample from the n sample units, for large samples, $V(\bar{y}_R) = (N - n_1)(S_y^2 + R^2 S_x^2 - 2RS_{xy})/Nn_1$.

If $\bar{X}$ is not known, it can be estimated by the mean $\bar{x}$ of the n units, provided there is complete response on the auxiliary characteristic. In this case, $\bar{Y}$ can be estimated from $\bar{y}_r = \hat{R}_1\bar{x}$. With the assumption of random responses, for large samples, the variance of this estimator is given by $(N - n)S_y^2/Nn + (n - n_1)(S_y^2 + R^2S_x^2 - 2RS_{xy})/nn_1$.

The large sample variances of both $\bar{y}_R$ and $\bar{y}_r$ are smaller than the variance $(N - n_1)S_y^2/Nn_1$ of $\bar{y}_1$, provided x and y are highly correlated. These variances can be estimated by replacing (S_x^2, S_y^2, S_{xy}) with $(s_{x1}^2, s_{y1}^2, s_{xy1})$ obtained from the n_1 responses and R by $\hat{R}_1$.

Notice that the sample mean can be expressed as $\bar{y} = (n_1/n)\bar{y}_1 + (n_2/n)\bar{y}_2$. When there is nonresponse on the n_2 units, if $\bar{y}_2$ is replaced by the predicted value $(\bar{y}_1/\bar{x}_1)\bar{x}_2$ obtained in Section 11.7, the estimator for the population mean now takes the form of $\bar{y}_r$. This estimator was also obtained by Jackson and P.S.R.S. Rao (1983) with suitable assumptions regarding the nonrespondents.

Example 11.4. Survey of families: The mean and variance of the 12 responses on income are $\bar{y}_1 = 47.5$ and $s_{y1}^2 = 256.73$. Hence, $v(\bar{y}_1) = (2000 - 12)(256.73)/2000(12) = 21.27$ and S.E.$(\bar{y}_1) = 4.61$. For the corresponding responses on husband's age, the mean and variance are $\bar{h}_1 = 36.75$ and $s_{h1}^2 = 83.48$, and its covariance with income is $s_{hy1} = 105.46$.

If the population mean for husband's age is 35, the ratio estimate for the average income is $(47.5/36.75)(35) = 45.2$. With the sample means and variances, an estimate of $V(\bar{y}_R)$ becomes $v(\bar{y}_R) = 10.24$, and hence S.E.$(\bar{y}_R) = 3.2$.

The mean of husband's age for the 16 sampled units is $\bar{h} = 38$. Now, the ratio estimator for the average income is $\bar{y}_r = (47.5/36.75)(38) = 49.1$. For this estimator, $v(\bar{y}_r) = 18.49$ and hence S.E.$(\bar{y}_r) = 4.3$.

In this illustration, for the average income, the sample mean 47.5 differs only a little from the ratio estimates 45.2 and 49.1. Also, the S.E. of $\bar{y}_R$ is only about two thirds of the S.E. of $\bar{y}_r$. Further, the decrease in the S.E. of $\bar{y}_r$ from $\bar{y}_1$ is not significant, since the population mean of husband's age is estimated from the sample.

Regression estimation

With the data from the n_1 respondents, a linear regression estimator for $\bar{Y}$ is given by $\bar{y}_L = \bar{y}_1 + b_1(\bar{X} - \bar{x}_1)$, where $b_1 = s_{xy1}/s_{x1}^2$. If the n_1 responses are considered to be random as before, for large samples, $V(\bar{y}_L) = (N - n_1)S_y^2(1 - \rho^2)/Nn_1$ This variance can be estimated by replacing S_y^2 with s_{y1}^2 and $S_y^2(1 - \rho^2)$ with $s_{e1}^2 = s_{y1}^2 - b_1^2s_{x1}^2$ obtained from the n_1 observations.

If $\bar{X}$ is not known, the regression estimator becomes $\bar{y}_l = \bar{y}_1 + b_1(\bar{x} - \bar{x}_1)$. The variance of this estimator is approximately given by $V(\bar{y}_l) = (N - n)S_y^2/Nn + (n - n_1)S_y^2(1 - \rho^2)/nn_1$. This variance can also be estimated as above.

11.9 Poststratification and weighting

Estimators for the mean and total

If the n units of the entire sample are poststratified into G strata, n_g of the units will be observed in the gth stratum. If the sizes N_g of the strata are known, an estimator for $\bar{Y}$ is

$$\hat{\bar{Y}}_W = \sum_1^G W_g \bar{y}_{g1}, \tag{11.20}$$

where $\bar{y}_{g1}$ is the mean of the n_{g1} respondents in the gth stratum and $W_g = N_g/N$.

Since $\bar{y}_{g1}$ is unbiased for the mean $\bar{Y}_{g1}$ of the N_{g1} respondents of the gth stratum, the bias of the above estimator is

$$B(\hat{\bar{Y}}_w) = \sum_1^G W_g \bar{Y}_{g1} - \sum_1^G W_g \bar{Y}_g = \sum_1^G \frac{N_{g2}}{N}(\bar{Y}_{g1} - \bar{Y}_{g2}) \tag{11.21}$$

In this expression, $N_{g2} = N_g - N_{g1}$ is the size of the nonrespondents in the gth stratum and $\bar{Y}_{g2}$ is their mean. If (N_{g2}/N) is small and $\bar{Y}_{g2}$ does not differ much from $\bar{Y}_{g1}$, as expected, this bias will be small.

For a given n_{g1}, the variance of $\hat{\bar{Y}}_W$ is

$$V(\hat{\bar{Y}}_W) = \sum_g W_g^2\left(\frac{1}{n_{g1}} - \frac{1}{N_{g1}}\right)S_{g1}^2. \tag{11.22}$$

The estimator of this variance is obtained by replacing S_{g1}^2 by the sample variance s_{g1}^2 of the n_{g1} respondents and N_{g1} by its estimate $N_g(n_{g1}/n_g)$.

If it is assumed that the n_{g1} respondents are a random sample of the n_g units, N_{g1} and S_{g1}^2 in the above expression should be replaced by N_g and S_g^2.

The estimator for the total is

$$\hat{Y}_W = \sum_1^G N_g \bar{y}_{g1} = \sum_1^G \frac{N_g}{n_{g1}} \sum_1^{n_{g1}} y_{gi}, \tag{11.23}$$

and its variance is obtained by multiplying (11.22) by N^2.

For the estimator in (11.20), the means of the respondents are inflated with the strata weights W_g. If they are not known, one can replace them by $w_g = n_g/n$. The variance in (11.22) can now be estimated by replacing N_{g1} and S_{g1}^2 by $N(n_{g1}/n)$ and s_{g1}^2 respectively. For this case of replacing W_g by w_g, Oh and Scheuren (1983) derive an approximation to the variance of the estimator in (11.20) with the assumption that the n_{g1} units respond with probabilities Q_g.

If the strata or *adjustment cells* are formed through a row × column classification, an estimator for the mean is $\Sigma_i \Sigma_j N_{ij}\bar{y}_{ij1}/N$, where $i = (1,...,r)$ and $j = (1,...,c)$ represent the rows and columns, respectively. In this expression N_{ij} is the number of units in the ijth cell and $\bar{y}_{ij1}$ is the mean of the n_{ij1} respondents in that cell in the sample of n units.

When a sample is classified as above, the proportions (N_{ij}/N) can be estimated from the sample proportion (n_{ij}/n). If the totals of the numbers of observations in the rows and in the columns, the marginal totals, are known, improvements on these estimates can be made through the **raking** method described, for example, by Brackstone and J.N.K. Rao (1976). For the case of nonresponse, Oh and Scheuren (1987) consider a modification of this procedure. Binder and Theberge (1988) derive the variance of an estimator obtained through the raking method.

For the estimation of the population proportion P of an attribute, the means in (11.20) and (11.21) are replaced by the corresponding proportions. The following example illustrates the effects of poststratification.

Example 11.5. Bias reduction through poststratification: Consider three strata for classifying the responses of a sample selected from a population of $N = 2000$ units. The sizes (N_{g1}, N_{g2}) of the respondents and nonrespondents and the numbers (C_{g1}, C_{g2}) of the units with an attribute of interest are presented in Table 11.2 for two compositions of the strata.

For the first type, the sizes of the three strata N_g are 900, 700, and 400. The numbers of units with the attribute of interest C_g are 510, 280, and

Table 11.2. Two types of stratification.

Sizes and Numbers	Type 1			Type 2		
N_{g1}	600	400	200	800	200	200
C_{g1}	360	160	20	200	180	160
N_{g2}	300	300	200	500	200	100
C_{g2}	150	120	10	120	80	80

30, respectively. Thus, the population proportion having the attribute is $P = (510 + 280 + 30)/2000 = 0.41$. For the second type, $N_g = (1300, 400, 300)$ and $C_g = (320, 260, 240)$.

The proportion for the 1200 respondents with the attribute is $P_1 = 540/1200 = 0.45$. Thus, if one does not poststratify the responses, the sample proportion p_1 of the n_1 responses has a bias of $0.45 - 0.41 = 0.04$.

For the respondents of the first type of stratification, the proportions with the attribute are $360/600 = 0.6$, $160/400 = 0.4$, and $20/200 = 0.10$. Thus, $\Sigma_1^G W_g P_{g1} = [9(0.6) + 7(0.4) + 4(0.1)]/20 = 0.43$. The bias of $\Sigma_1^G W_g P_{g1}$ for estimating the population proportion is $0.43 - 0.41 = 0.02$, which is only half the bias for p_1.

For the second type of stratification, the proportions with the attribute are $200/800 = 0.25$, $180/200 = 0.9$, and $160/200 = 0.8$, and hence $\Sigma_1^G W_g P_{g1} = [13(0.25) + 4(0.9) + 3(0.8)]/20 = 0.4625$. The bias of $\Sigma_1^G W_g P_{g1}$ now is $0.4625 - 0.41 = 0.0525$, which is larger than the bias of both p_1 and the estimator with the first type of stratification.

As seen in the above example, for a particular characteristic, one type of stratification can be more beneficial than the other. Proper poststratification followed by ratio or regression methods of estimation can be helpful in reducing the bias and MSE of the estimators based on the respondents.

Bailar et al. (1978) describe the procedures used for adjusting for the noninterviews in the Current Population Survey (CPS). Six cells based on three regions and two color categories were used for stratification. For estimating income-related variables in the CPS, Ernst (1978) examined the effects of weighting classes based on characteristics such as age, race, sex, educational level, occupational characteristics, and marital status of the head of the family.

The effects of several weighting procedures on the observations of a national telephone survey on smoking and health-related characteristics were evaluated by Boteman et al. (1982). The weights were based on the probability of selection of the units, nonresponse, telephone coverage, and poststratification. Jagers (1986) describes the procedures for reducing the nonresponse bias through poststratification.

Subpopulations

As in Section 6.7, let N_{gj}, Y_{gj}, $\bar{Y}_{gj}$, and S_{gj}^2, $g = (1, ..., G)$ and $j = (1, ..., k)$, denote the size, total, mean, and variance of the jth subpopulation in the gth stratum. As before, let the additional subscripts 1 and 2 denote the respondents and nonrespondents.

If N_{gj} is known, an estimator for the population total $Y_j = \Sigma_g Y_{gj}$ of the $N_j = \Sigma_g N_{gj}$ units of the subpopulation is

$$\hat{Y}_j = \sum_1^G \hat{Y}_{gj} = \sum_1^G N_{gj} \bar{y}_{gj1}, \tag{11.24}$$

where $\bar{y}_{gj1}$ is the mean of the n_{gj1} respondents among the n_{gj} units in the gth stratum.

Since $\bar{y}_{gj1}$ is unbiased for $\bar{Y}_{gj1}$, the bias of this estimator is

$$B(\hat{Y}_j) = \sum_1^G N_{gj} \bar{Y}_{gj1} - \sum_1^G N_{gj} \bar{Y}_{gj} = \sum_1^G N_{gj2}(\bar{Y}_{gj1} - \bar{Y}_{gj2}). \tag{11.25}$$

Its variance is given by

$$V(\hat{Y}_j) = \sum_1^G N_{gj}^2 \left(\frac{1}{n_{gj1}} - \frac{1}{N_{gj1}} \right) S_{gj1}^2. \tag{11.26}$$

For estimating this variance, replace S_{gj1}^2 by the variance s_{gj1}^2 of the n_{gj1} respondents and N_{gj1} by its estimate $N_{gj}(n_{gj1}/n_{gj})$.

The estimator for the mean $\bar{Y}_j$ of the jth domain is given by $\hat{Y}_j/N_j$. If N_{gj} is not known, but N_g is known, for estimating the total and mean, replace it by its estimate $N_g(n_{gj}/n_g)$. If N_g is not known, it is estimated from $N(n_g/n)$ and hence N_{gj} is estimated from $N(n_{gj}/n)$. The following example will illustrate this procedure.

Example 11.6. Weighting for subpopulations: In a random sample of 400 from the 3600 undergraduate students of a university, 280 responded to a question on the number of hours of part-time work during a week. The results were poststratified into male and female groups. Figures for the freshman–sophomore (FS) and junior–senior (JS) classes are presented in Table 11.3.

With the assumption that all the four groups are of the same size, $\hat{N}_{gj}$ = 900. Now, an estimate of the average for the FS class is $\hat{\bar{Y}}_j = 0.5(8.5) +$

Table 11.3.   Part-time employment of students.

	Male		Female	
	FS	JS	FS	JS
Sample size	55	165	40	140
No. of responses	45	120	30	85
Mean	8.5	12.5	10.3	14.4
Variance	24	22	26	32

$0.5(10.3) = 9.4$. An estimate of the variance for this average becomes

$$v(\hat{\bar{Y}}_j) = 0.25 \frac{(900 - 45)}{900(45)} (24) + 0.25 \frac{(900 - 30)}{900(30)} (26) = 0.336$$

and hence it has a S.E. of 0.58.

If N_{gj} are estimated from $N(n_{gj}/n)$, the sizes for the male and female FS class are $3600(55/400) = 495$ and $3600(40/400) = 360$. Thus, they are in the proportion $495/855 = 0.58$ and 0.42. The estimate for the average for the FS class now is given by $\hat{\bar{Y}}_j = 0.58(8.5) + 0.42(10.3) = 9.256$. The estimate of variance in this case is

$$v(\hat{\bar{Y}}_j) = (0.58)^2 \frac{(495 - 45)}{495(45)}(24) + (0.42)^2 \frac{(360 - 30)}{360(30)}(26) = 0.3032$$

and hence it has a S.E. of 0.55.

For the case of unknown domain sizes, Little (1986) compared the mean of the respondents $\Sigma_g n_{gj1} \bar{y}_{gj1} / \Sigma_g n_{gj1}$ with the *weighted in cell* mean $\Sigma_g w_{gj1} \bar{y}_{gj1} / \Sigma_g w_{gj1}$, where $w_{gj1} = (n_g/n_{g1})n_{gj1}$.

11.10   Response probabilities and weighting

As noted in Section 2.10, for the case of simple random sampling and complete response, the estimator $N\bar{y}$ for the total can be expressed as $\Sigma^n (y_i/\phi_i)$ where $\phi_i = (n/N)$ is the probability of selecting a unit into the sample. With stratification, the estimator for the population total $\hat{Y}_{st} = N\bar{Y}_{st}$ studied in Chapter 5 can be expressed as $\Sigma_g(t_g/\phi_g)$, where $t_g = \Sigma^{n_g} y_{gi}$ is the sample total and $\phi_g = n_g/N_g$.

In the case of incomplete response, the sampled units can be considered to respond with certain probabilities, which can be estimated from the sample. Estimators for the population mean with weights based on the probabilities of responses are described below. The biases and MSEs of these estimators can be evaluated for suitable response probabilities.

For the case of a simple random sample of n units selected without replacement and n_1 responses, an estimate for $\bar{Y}$ can be expressed as

$$\hat{\bar{Y}}_Q = \frac{\sum^{n_1} (y_i/Q_i)}{\sum^{n_1} (1/Q_i)}, \tag{11.27}$$

where Q_i is the conditional probability of response of the ith unit which was selected into the sample. If $Q_i = n_1/n$, this estimator is the same as $\bar{y}_1$.

If the n sample units are selected with unequal probabilities π_i, the Horvitz–Thompson type estimator with the n_1 responses can be expressed as

$$\hat{\bar{Y}}_{HQ} = \frac{\sum^{n_1}(y_i/\pi_i Q_i)}{\sum^{n_1}(1/\pi_i Q_i)}. \tag{11.28}$$

If $\pi_i = n/N$ and Q_i are all equal, this estimator again is the same as $\bar{y}_1$.

With nonresponse and poststratification, the estimator in (11.27) can be expressed as

$$\hat{\bar{Y}}_{SQ} = \frac{\sum_g \sum_i (y_{gi}/\pi_{gi} Q_{gi})}{\sum_g \sum_i (1/\pi_{gi} Q_{gi})}, \tag{11.29}$$

where π_{gi} is the probability of selecting a unit into the gth stratum and Q_{gi} is its probability of response. In this expression, the second summation is carried over the number of responses n_{g1} in the gth stratum. Special cases of this estimator are as follows.

If $\pi_{gi} = n_g/N_g$ and $Q_{gi} = n_{g1}/n_g$, the above estimator becomes the same as the poststratified mean in (11.20). In this case, the response probabilities for the units of the gth stratum are estimated from n_{g1}/n_g.

If $\pi_{gi} = n/N$ and $Q_{gi} = Q_g$, (11.29) becomes

$$\hat{\bar{Y}}_{SQ} = \frac{\sum n_{g1}\bar{y}_{g1}/Q_g}{\sum n_{g1}/Q_g}. \tag{11.29a}$$

The estimator $\bar{y}_H$ in (11.2) for subsampling the nonrespondents can be obtained from this expression by replacing (n_{11}, n_{21}) with (n_1, m), $(\bar{y}_{g1}, \bar{y}_{g2})$ with $(\bar{y}_1, \bar{y}_{2(m)})$, and noting that $Q_1 = 1$ and $Q_2 = m/n_2$.

A procedure for obtaining an estimator of the type in (11.29a) was considered by Politz and Simmons (1949) and Simmons (1954). In this method, interviews were assumed to be conducted during the six evenings from Monday through Saturday. The number of evenings the respondent was home during the previous five evenings was recorded at the time of the interview. If the respondent was home on g evenings, $g = (0,...,5)$, an estimate of the probability of his or her response is $(g + 1)/6$. With this estimate, the population mean is obtained from (11.29a).

If the nonresponse of a unit on a characteristic y_i depends on an auxiliary or supplementary variable x_i, Little and Rubin (1987) characterize it as the *ignorable nonresponse*. In this case, the probability of nonresponse can be estimated from the available information, and the above type of weighting procedures can be employed. If the non-response cannot be completely explained by x_i, it is recognized as *nonignorable nonresponse*. Suitable procedures are required in this case to adjust the estimators for the nonresponse. If neither of these two cases can explain the nonresponse, in some situations the responses and nonresponses may be considered to be obtained from random samples of the sampled units.

For the case of Q_i depending on an auxiliary characteristic, the merits of the resulting estimator relative to $\bar{y}_1$ were examined by Oh and Scheuren (1983) and Little (1986). Cassell et al. (1983) suggest the estimation of Q_i from the available auxiliary information.

11.11   Imputation

For some large-scale surveys, the effects of imputing suitable values for the missing units have been examined. If the imputation is preceded by grouping the responding units into homogeneous strata or cells, it can help reduce the nonresponse bias in some cases. Effects of some of the imputation schemes can be similar to the regression and weighting adjustments described in the previous sections. Another reason given for imputing the missing values is that the users of surveys may find it convenient to analyze the *clean* set of data obtained after imputation than the original data with observations missing on different characteristics for different units.

Mean imputation

In this procedure, the mean $\bar{y}_1$ of the n_1 respondents are duplicated for the $n_2 = n - n_1$ nonrespondents. The resulting estimator $\bar{y}_d$, with the subscript d denoting duplication, remains the same as $\bar{y}_1$, and its variance is the same as $(N_1 - n_1)S_1^2/n_1$. As noted earlier, the sample variance s_1^2 of the n_1 units is unbiased for S_1^2. With the completed observations, the sample variance becomes

$$s_d^2 = \frac{\sum_1^n (y_i - \bar{y}_d)^2}{n-1} = \frac{n_1 - 1}{n-1} s_1^2. \tag{11.30}$$

Hence, s_d^2 underestimates s_1^2. Further, the distribution of the n completed observations is concentrated around $\bar{y}_1$ and does not properly represent the frequency distribution of the n_1 responses.

Random duplication

For this method, n_2 values selected randomly without replacement from the n_1 respondents are imputed for the nonrespondents. In the completed sample, n_2 observations appear twice and the remaining $n - 2n_2 = n_1 - n_2$ observations appear once. Denote the duplicated observations by y_{i0}, $i = (1,...,n_2)$, and let $\bar{y}_0$ and s_0^2 denote their mean and variance.

With the completed observations, an estimator for the population mean is

$$\bar{y}_c = \frac{\sum^{n_1} y_i + \sum^{n_2} y_{i0}}{n} = w_1 \bar{y}_1 + w_2 \bar{y}_0, \tag{11.31}$$

where $w_1 = n_1/n$ and $w_2 = n_2/n$ as before. Let I denote the initial sample and the responses. Since $E(\bar{y}_0 | I) = \bar{y}_1$, $E(\bar{y}_c) = \bar{y}_1$. Hence, the bias of $\bar{y}_c$ remains the same as that of $\bar{y}_1$. Following Appendix A3,

$$V(\bar{y}_c) = V(\bar{y}_1) + w_2^2 \left(\frac{n_1 - n_2}{n_1 n_2} \right) S_1^2$$

$$= \left[\frac{1}{n}(1 + 2w_2) - \frac{1}{N_1} \right] S_1^2. \tag{11.32}$$

For large N_1, from (11.32), the proportional increase in the variance is

$$\frac{V(\bar{y}_c) - V(\bar{y}_1)}{V(\bar{y}_1)} = w_2^2 \frac{n_1 - n_2}{n_2} = w_2(1 - 2w_2). \tag{11.33}$$

This expression reaches its maximum at $w_2 = 1/4$. In this case, the relative increase in the variance due to duplication is 12.5%.

Variance estimation

An unbiased estimator of (11.32) is obtained by replacing S_1^2 with s_1^2. The variance s_c^2 of the completed observations can be expressed as

$$(n - 1)s_c^2 = (n_1 - 1)s_1^2 + (n_2 - 1)s_0^2 + nw_1w_2(\bar{y}_0 - \bar{y}_1)^2. \quad (11.34)$$

Since the conditional expectations of s_0^2 and $(\bar{y}_0 - \bar{y}_1)^2$ are s_1^2 and $(n_1 - n_2)s_1^2/n_1n_2$, from the above expression,

$$E(s_c^2|I) = \frac{n(n - 2) + (n_1 - n_2)}{n(n - 1)}s_1^2 = \left[1 - 2\frac{w_2}{n - 1}\right]s_1^2 \quad (11.35)$$

Thus, s_c^2 underestimates s_1^2 and hence S_1^2. This underestimation, as expected, increases with the number of nonresponses but becomes negligible for large n. An unbiased estimator for S_1^2 is given by $(n - 1)s_c^2/(n - 1 - 2w_2)$.

Confidence limits

If the responses are considered to be a random sample from the n selected units, both $\bar{y}_1$ and $\bar{y}_c$ become unbiased for $\bar{Y}$. In this case, the variance of $\bar{y}_c$ is obtained by replacing N_1 and S_1^2 in (11.32) with N and S^2, respectively.

If the n_1 responses are considered to be a random sample from the n sampled units, for large N, the variance in (11.32) can be estimated from $(1 + 2w_2)s_1^2/n$. Now, $(1 - \alpha)\%$ confidence intervals for $\bar{Y}$ are obtained from

$$C_1 : \bar{y}_c \pm Z\left(\frac{1 + 2w_2}{n}\right)^{1/2}s_1 \quad (11.36)$$

Consider the alternative limits,

$$C_2 : \bar{y}_c \pm Zs_c/\sqrt{n}. \quad (11.37)$$

Note from (11.35) that the expected width of C_2 is smaller than the width of C_1.

Multiple imputation

Rubin (1978; 1979; 1986) suggests and illustrates this procedure for the estimation of means, totals, proportions, and other population quantities. One purpose of this procedure is to avoid underestimation of the variances and standard errors, noted in the above sections. Rubin (1987) presents the details of this approach, and its implementation is reviewed in Rubin (1996).

In this procedure, m sets of all the missing observations are randomly selected from a suitable *posterior distribution*. The derivation of a posterior distribution is briefly outlined in Appendix A12. With the selected observations of each set and the corresponding responses, estimates $\hat{T}_i$ for the population quantity and the variance v_i, $i = 1, 2, ..., m$ of the completed observations are obtained. The final estimate $\hat{T}$ is obtained from the average of the $\hat{T}_i$. The variance and the standard error of $\hat{T}$ are obtained by combining the variance among the $\hat{T}_i$ and all the within variances v_i.

This method is also illustrated in Herzog and Rubin (1983), Rubin and Schenker (1986), and Heitjan and Little (1988). Rubin et al. (1988) describe this approach for postenumeration surveys, and Glynn et al. (1993) for surveys with follow-ups. Gelman et al. (1995; 1998) describe multiple imputation and the Bayesian approach. Schafer and Schenker (2000) present a procedure for replacing the missing observations with predictions derived from a suitable model for imputation.

Hot- and cold-deck imputation

For large-scale surveys, these procedures were implemented by the U.S. Bureau of the Census and other organizations. For the *hot deck* method, observations are imputed from the current (*hot*) sample, with its units arranged in some order, for example, as a *deck* of computer cards. In contrast, for the *cold-deck* method, observations for imputation are selected from previous surveys, censuses, and similar sources.

Bailar et al. (1978) describe the hot-deck procedure used in the CPS to impute for nonresponse on the different variables; for complete nonresponse on the units, weighting procedures are used as described in Section 11.9. For imputation on the employment-related items, the sample units—about 50,000 households—were arranged in 20 cells. Five age groups, white and black or Hispanic classification, and male–female categories were used to form these cells. A sequential procedure of imputing observations from the most recently updated records was implemented. With this method, the same observation can be duplicated many times. For large sizes of the population and the sample, the variance of the imputed mean derived by the above authors, with the assumption that the responses constitute a random sample from the initial sample, is given by $(1 + 2n_2/n_1)S^2/n$.

11.12   Related topics

Efforts to increase response rates

For several types of surveys and polls, short and unambiguous questions were found to yield high response rates. They are usually low on personal questions, for example, on income and family finances. Response rates cannot be expected to be high on sensitive questions such as smoking habits or alcohol and drug addiction. To elicit responses on such items by assuring confidentiality, Warner (1965) and others suggest the **randomized response** method. Sirken (1983) describes **network sampling** to contact and obtain responses from the sample units. To enumerate the homeless and transient, Iachan and Dennis (1993) present the multiple frame methodology.

Imputation, estimation, and analysis

Bailar and Bailar (1978; 1983) compare the biases arising from the hot-deck and other types of imputation. Ernst (1980) derives the variances for the estimators of the mean obtained through different types of imputation. Chapman et al. (1986) summarize the imputation methods implemented by the U.S. Bureau of the Census. Kalton and Kish (1984) and Kalton and Kasprzyk (1986) describe the different procedures of imputation used in practice. Fellegi and Holt (1976), Platek and Gray (1983), Giles (1988), for example, describe the different types of imputation for the surveys conducted by Statistics Canada. Fay (1996) describes some of the procedures for imputing for the nonrespondents.

Analysis and estimation from imputed data are examined, for example, by Titterington and Sedransk (1986), Wang et al. (1992), and Kott (1994a). J.N.K. Rao and Shao (1992) and J.N.K. Rao (1996) present procedures for finding the S.E. of the estimates obtained from imputed data.

Models for prediction, poststratification, and weighting

Regression type of models for the prediction of observations for the non-respondents or poststratification followed by weighting were suggested by Little (1982; 1995), Särndal (1986), Rancourt et al. (1994), Kott (1994b), and others. For longitudinal data, Stasny (1986; 1987), Duncan and Kalton (1987), Nordberg (1989), and Little (1995), for example, describe estimation for the case of nonresponse. Potthoff et al. (1993) describe weighting of the responses obtained from the callbacks.

Qualitative and categorical observations

When there is complete response, as in Chapter 4, denote by p the sample proportion having an attribute of interest, and $q = 1 - p$ its complement. The transformed variable $y = \log_e(p/q)$, where $\log_e$ stands for the logarithm with base e, is the *logit* of p. In some situations, this transformed variable is assumed to follow the regression model of the type in (10.5) and its extensions. To adjust for the nonresponse, this type of *logistic regression* is considered, for example, by Alho (1990).

If the characteristic is of the low–medium–high or similar categorical type, the proportions for the different classes can be transformed as above, and adjustments for the nonresponses can be considered. Fay (1986), Baker and Laird (1988), Binder (1991), Conaway (1992), Lipsitz et al. (1994), and others examine such procedures.

11.13   Bayesian procedures

As outlined briefly in Appendix A12, in these procedures, formalized prior beliefs, and information are combined with the sample information. Ericson (1969), Malec and Sedransk (1985), Calvin and Sedransk (1991), for example, describe this approach for finite population sampling. For the case of nonresponse, the Bayesian method is presented by Ericson (1967), J.N.K. Rao and Ganghurde (1972), Smouse (1982), Kadane (1993), and others. To adjust the nonresponse of categorical data, Bayesian procedures were described, for example, by Kaufmann and King (1973), Chiu and Sedransk (1986), and Raghunathan and Grizzle (1995).

11.14   Cesnsus undercount

Following the 1980 Decennial Census of the U.S., adjustments for the undercount of the population in several areas were examined through imputation, poststratification, weighting, and other procedures described in the above sections. Ericsen and Kadane (1985) and Ericsen et al. (1992) considered suitable models for this purpose. Cressie (1989; 1992) and others examine the Bayesian approach.

Fienberg (1992) provides bibliography for the capture–recapture method described in Section 4.12 and its application for the census undercount. Hogan (1993) and Ding and Fienberg (1994) describe the related *dual system* estimation for combining information from two or more sources, such as the census and the post-censal surveys.

Exercises

11.1. For the survey described in Example 11.1, consider $n_1 = 250$ responses for another characteristic. If 150 of the respondents are observed to have the attribute of interest, find the 95% confidence limits for the percentage of the population having the attribute.

11.2. If $e_0 = 0.5$, $e_1 = 1$, $e_2 = 6$, and $6000 are available for the survey in Example 11.2, find the optimum sizes for the samples at the two stages. If $S^2 = 2500$, find the S.E. of $\bar{y}$ that is obtained with these optimum sizes.

11.3. Consider the eight families in Table 11.1 with the completed observations on income, husband's age, and wife's age. Fit the multiple regressions of income on the ages of the couples through the origin and with the intercept, and test for the significance of the regressions. To predict the missing observations on income, examine whether these regressions can be preferred to the linear regressions on husband's and wife's age analyzed in Example 11.3.

11.4. From the responses of the families in Table 11.1, fit the linear regressions of television time on each of the remaining four variables, and examine which of the four regressions can be recommended for predicting the missing observations on television time.

11.5. In a sample of 150 from the 2000 population units of Example 11.5, only 95 responded. The observed sample sizes for the two types of stratification are (42, 35, 18) and (55, 26, 14). To estimate the population proportion, for both these cases, compare the variances and MSEs of the sample proportion and the poststratified estimator of the proportion (a) assuming that the responses are a random sample and (b) without this assumption.

11.6. Among the 95 respondents of the above example, 46 were observed to have the characteristic of interest. The numbers observed in the three strata of the first type were (22, 15, 9). (a) Compare the sample estimate of the proportion for the characteristic of interest with the poststratified estimate. (b) With the assumption of random response, find the sample standard errors of these estimates.

11.7.   (a) Using the data in Table 11.3, find estimates for the average number of hours of part-time employment for the junior–senior class through the two procedures described in Example 11.6 and find their S.E. values.

11.8.   If $n_1 < n_2$, the n_2 observations can be duplicated with replacement from the n_1 responses. (a) Show that the expectation of $\bar{y}_c$ for this case is also $\bar{y}_1$, but its variance is obtained by replacing $(n_1 - n_2)$ in the second term of (11.32) by $(n_1 - 1)$. (b) Show that the increase in the variance of $\bar{y}_c$ relative to $\bar{y}_1$ for this type of duplication approximately becomes $w_1 w_2$, and it reaches its maximum of 25% when the response rate is 50%.

Appendix A11

Variance of the Hansen–Hurwitz estimator

First note that $V[E(\bar{y}_H \,|\, s)] = V(\bar{y}) = (1 - f)S^2/n$, where $f = n/N$ and S^2 is the variance of the N population units. Next, from (11.2),

$$V(\bar{y}_H \,|\, s; n_2) = w_2^2 \frac{k-1}{n_2} s_2^2 = w_2 \frac{k-1}{n} s_2^2,$$

where s_2^2 is the variance of the n_2 units of the second stratum. Since $E(s_2^2) = S_2^2$, which is the variance of the N_2 nonresponding units, and $E(w_2) = W_2 = N_2/N$,

$$E[V(\bar{y}_H \,|\, s; n_2)] = W_2 \frac{k-1}{n} S_2^2.$$

The variance in (11.3) is obtained by combining these two expressions.

Variance estimator for the Hansen–Hurwitz estimator

The sample variance s^2 of the n units is unbiased for S^2, but it is not available due to the nonresponse. From the results of the samples at the two stages, an unbiased estimator of S^2 is given by

$$\hat{S}^2 = \frac{n_1 - 1}{n - 1}s_1^2 + \frac{n_2 - k + w_2(k - 1)}{n - 1}s_m^2$$

$$+ \frac{n}{n - 1}[w_1(\bar{y}_1 - \bar{y}_H)^2 + w_2(\bar{y}_m - \bar{y}_H)^2].$$

The term in the square brackets is the same as $w_1 w_2(\bar{y}_1 - \bar{y}_m)^2$.

Replacing S^2 by the above estimator and S_2^2 by s_m^2, an unbiased estimator for the variance in (11.3) is obtained from

$$v(\bar{y}_H) = \frac{(1-f)}{n(n - 1)}[(n_1 - 1)s_1^2 + (n_2 - k)s_m^2 + n_1(\bar{y}_1 - \bar{y}_H)^2$$

$$+ n_2(\bar{y}_m - \bar{y}_H)^2] + \frac{(N - 1)(k - 1)w_2}{N(n - 1)}s_m^2.$$

CHAPTER 12

Further Topics

12.1 Introduction

This chapter contains three major topics of interest in sample surveys. The first topic in Sections 12.2 through 12.5, presents the linearization, jackknife, bootstrap, and balanced repeated replication procedures. The last three methods in this group are also known as the **resampling procedures**. Estimating the variances and MSEs of nonlinear estimators, such as the ratio of two sample means and also more-complicated estimators, and finding confidence limits for the corresponding population quantities are two of the major purposes of these approaches. The linearization and jackknife procedures can also be used to reduce the bias of an estimator.

The second topic, presented in Section 12.6, describes the different procedures available for estimation of the population quantities for **small areas**, domains, and subpopulations of small sizes. The final section describes estimation methods for **complex surveys**.

12.2 Linearization

For the sake of illustration, consider estimation of the ratio $R = \overline{Y}/\overline{X}$ of two means, as in Chapter 9. As described in Appendix A9, the large sample variance of $\hat{R} = \bar{y}/\bar{x}$ in (9.1) was found by expanding $(\hat{R} - R)$ in a series and ignoring higher-order terms, that is, through the **linearization** of $\hat{R}$. The estimator $v(\hat{R})$ for this variance was also found using this procedure and was presented in (9.2).

Reducing the bias of $\hat{R}$ through linearization, the estimator $\hat{R}_T = \hat{R}[1 - (1 - f)(c_{xx} - c_{xy})/n]$ in (9.24) was obtained. Thus, linearization can be employed for finding an approximate expression for the variance or MSE of a nonlinear estimator, to estimate it from the sample and also to reduce its bias. The following example compares $\hat{R}$ and $\hat{R}_T$.

Example 12.1. College enrollments: For 1990, enrollment for private colleges (y_i) and the total enrollment for private and public colleges (x_i) appear in the *Statistical Abstracts of the United States* (1992), Table 264. For the $N = 48$ states, excluding the two largest states, California and New York, the ratio of the means of the private and total enrollment is $R = 0.2092$.

These enrollments for a sample of $n = 10$ from the 48 states are presented in the second and third columns of Table 12.1, along with their means and variances. From these figures, $\hat{R} = 52.8/198.7 = 0.2657$. Further, the sample covariance is $s_{xy} = 6949.4$. Now, $s_d^2 = s_y^2 + \hat{R}^2 s_x^2 - 2\hat{R} s_{xy} = 4508.6 + (0.2657)^2(17,628) - 2(0.2657)(6949.4) = 2060.14$. Thus, from (9.2), $v(\hat{R}) = [(48 - 10)/480](2060.14)/(198.7)^2 = 0.0041$.

Since the sample mean and variance of x are 198.7 and 17,628, $c_{xx} = 17,628/198.7^2 = 0.4465$. Since the sample standard deviations of x and y are 132.8 and 67.1, and the sample covariance is 6949.4, $c_{xy} = 6949.4/(198.7)(52.8) = 0.6624$.

From the above figures, $\hat{R}_T = 0.2705$, which does not differ much from $\hat{R} = 0.2657$. Since the sample is nearly 20% of the population, the reduction in the bias is not large.

In general, linearization conveniently provides expressions for the biases and variances of complicated estimators and it is routinely employed in theoretical and applied statistics. As another example, through this approach, Shao and Steel (1999) develop a variance estimator for the Horvitz–Thompson estimator in (7.34) obtained with imputed data. Woodruff (1971) notes that approximations to the variance of an estimator and its estimate can also be obtained through the *Taylor's series expansion*. Wolter (1984) describes linearization and other methods for variance estimation.

12.3   The jackknife

Quenouille (1956) proposed a method for reducing the bias of an estimator. Since it can also be used for estimating variances, finding confidence limits, and similar purposes, Tukey (1958) popularized it as the *jackknife*, a pocket- or penknife. The following subsections present this procedure for $\hat{R}$.

Bias reduction

For a large population, the expectation of $\hat{R}$ can be expressed as

$$E(\hat{R}) = R + c_1/n + c_2/n^2 + ..., \tag{12.1}$$

where $c_1, c_2, c_3, ...$ are the expectations of moments of increasing degree.

If the ith pair of the observations, $i = 1, 2, ..., n$, is deleted from the sample, the means become $\bar{y}_{n-1,i} = (n\bar{y} - y_i)/(n - 1)$ and $\bar{x}_{n-1,i} = (n\bar{x} - x_i)/(n - 1)$. As in the case of $\hat{R}$, the expectation of the ratio $\hat{R}_{n-1,i} = \bar{y}_{n-1,i} / \bar{x}_{n-1,i}$ can be expressed as

$$E(\hat{R}_{n-1,i}) = R + c_1/(n - 1) + c_2/(n - 1)^2 + ... \qquad (12.2)$$

Consider the *pseudo* values

$$\hat{R}_i' = n\hat{R} - (n - 1)\hat{R}_{n-1,i} \qquad (12.3)$$

for $i = 1, 2, ..., n$. From (12.1) and (12.2),

$$E(\hat{R}_i') = R - c_2/n(n - 1) - (2n - 1)c_3/n^2(n - 1)^2 + -.... \qquad (12.4)$$

The biases of these pseudo values are approximately of order $1/n^2$. The jackknife estimator for R is given by their average,

$$\hat{R}_J = \sum \hat{R}_i'/n = n\hat{R} - (n - 1)\hat{R}_{n-1}, \qquad (12.5)$$

where $\hat{R}_{n-1} = \Sigma \hat{R}_{n-1,i}/n$, and from (12.4) its bias is also approximately of order $1/n^2$.

In the original development of this procedure, Quenouille divides the initial sample into g groups of sizes $m = n/g$ each, deletes one group at a time, constructs the pseudo values, and considers their average as the estimator. For example, when the sample is divided into two groups of size $m = n/2$ each, denote their means by $(\bar{x}_1, \bar{y}_1)$ and $(\bar{x}_2, \bar{y}_2)$, with the corresponding ratios $\hat{R}_1 = \bar{y}_1 / \bar{x}_1$ and $\hat{R}_2 = \bar{y}_2 / \bar{x}_2$ The pseudo values now are $2\hat{R} - \hat{R}_2$ and $2\hat{R} - \hat{R}_1$, and the jackknife estimator for R is given by $\hat{R}_J = 2\hat{R} - (\hat{R}_1 + \hat{R}_2)/2$.

Variance estimation

For sampling from a finite population, the jackknife estimator suggested for the MSE of $\hat{R}_J$ or $\hat{R}$ is obtained from

$$vj(\hat{R}) = [(1 - f)/n]\sum (\hat{R}_i' - \hat{R}_J)^2/(n - 1)$$

$$= (1 - f)(n - 1)\sum (\hat{R}_{n-1,i} - \hat{R}_{n-1})^2/n \qquad (12.6)$$

Table 12.1. Jackknife calculations for college enrollment.

Sample No.	Total Enrollment (x)	Private Enrollment (y)	$\bar{x}_{n-1,i}$	$\bar{y}_{n-1,i}$	$\hat{R}_{n-1,i}$
1	90	12	210.78	57.33	0.2720
2	227	26	195.56	55.78	0.2852
3	252	55	192.78	52.56	0.2726
4	171	53	201.78	52.78	0.2616
5	57	16	214.44	56.89	0.2653
6	419	234	174.22	32.67	0.1875
7	352	67	181.67	51.22	0.2820
8	34	8	217.00	57.78	0.2663
9	85	11	211.33	57.44	0.2718
10	300	46	187.44	53.56	0.2857
Average	198.7	52.8	198.7	52.8	0.2650
Variance	17,628.0	4508.6			0.000811

For the case of two groups, this estimator takes the form of $[(1 - f)/n]$ $(\hat{R}_1 - \hat{R}_2)^2/2$.

The estimator for the mean and its variance are obtained from $\bar{X}\hat{R}_J$ and $\bar{X}^2 v_J(\hat{R})$. For the college enrollments, the calculations needed for the jackknife are presented in Table 12.1. With these figures, from (12.5), $\hat{R}_J = 10(0.2657 - 9(0.265) = 0.272$. This estimate is not too far from $\hat{R} = 0.2657$ and $\hat{R}_T = 0.2705$. From (12.6), $v_J(\hat{R}) = (38/48)$ $(9/10)(0.007299) = 0.0052$.

Durbin (1959), J.N.K. Rao and Webster (1966), Rao and Rao (1969), P.S.R.S. Rao (1969; 1974), Krewski and Chakrabarti (1981), Kreswski and J.N.K. Rao (1981), and others evaluated the merits of the linearization and jackknife methods for reducing the bias of the ratio estimator and estimating its variance, through the model in (9.10) and suitable assumptions. As g increased to n, the bias and MSE of $\hat{R}_J$ were found to decrease. However, the variance of $v_J(\hat{R})$ was found to be larger than that of $v(\hat{R})$ in (9.2) obtained from the linearization procedure; that is, $v_J(\hat{R})$ can be less **stable**. Since the S.E. for an estimator is obtained from its variance estimator, stability is a desirable property.

The jackknife procedure for ratio estimation with stratification is presented in Jones (1974). Following this approach, for a finite population, n and $(n - 1)$ in (12.5) should be replaced by $w = n(N - n + 1)/N$ and $(1 - w) = (n - 1)(N - n)/N$. Schucany et al. (1971) consider higher-order jackknife by the reapplication of the procedure. Since the ratio estimator for the mean $\hat{R}\bar{X}$ can be expressed as $\bar{y} + \hat{R}(\bar{X} - \bar{x})$,

P.S.R.S. Rao (1979) considered $\bar{y} + \hat{R}_J(\bar{X} - \bar{x})$ for reducing its bias, and compared it with $\hat{R}_J\bar{X}$.

For variance estimation, J.N.K. Rao and Shao (1992) examine the jackknife for the data from hot-deck imputation, J.N.K. Rao (1996) for the ratio and regression methods of imputation in single and multi-stage sampling, and Yung and J.N.K. Rao (1996) for stratified multistage sampling.

12.4   The bootstrap

This method proposed by Efron (1982) and also presented in Efron and Tibshirani (1993) can be used for estimating the variance or MSE of a nonlinear estimator and for finding confidence limits.

Different procedures of applying the bootstrap for variance estimation and other purposes were suggested by Gross (1980), McCarthy and Snowden (1985), J.N.K. Rao and Wu (1987), Kovar et al. (1988), Sitter (1992), among others. Booth et al. (1994) examine the coverage probabilities of the confidence limits obtained from applying the bootstrap procedure to the separate and combined ratio estimators. Efron (1994) describes the bootstrap for missing data and imputation. Shao and Tu (1995) present it along with the jackknife method, and Shao and Sitter (1996) examine it for the imputed data. For the variance of the two-phase regression estimator, Sitter (1997) compares it with the linearization and jackknife methods. C.R. Rao, et al. (1997) suggest a procedure for selecting the bootstrap samples.

The bootstrap method is illustrated below for estimating the variance of the mean of a sample selected from an infinite or finite population and for the estimation of $V(\hat{R})$.

Infinite population

Consider an infinite population for the characteristic y with mean $\mu = E(y)$ and variance $\sigma^2 = E(y - \mu)^2$. The mean $\bar{y} = \Sigma_1^n y_i/n$ and variance $s^2 = \Sigma_1^n (y_i - \bar{y}^2)/(n - 1)$ of a sample of observations, y_i, $i = 1, ..., n$, are unbiased for μ and σ^2, respectively. The variance of $\bar{y}$ and its unbiased estimator are $V(\bar{y}) = \sigma^2/n$ and $v(\bar{y}) = s^2/n$.

Let $\bar{y}_b = \Sigma_1^m y_i'/m$ and $s_b^2 = \Sigma_1^m (y_i - \bar{y})^2/(m - 1)$ denote the mean and variance of the observations y_i', $i = 1, 2, ..., m$ of a sample of size m selected *with* replacement from the above sample. With I denoting the initial sample, $E(\bar{y}_b \mid I) = \bar{y}$ and

$$v(\bar{y}_b \mid I) = \frac{1}{m} \sum_1^n \frac{(y_i - \bar{y})^2}{n} = \frac{1}{m}\frac{(n-1)}{n}s^2. \tag{12.7}$$

If $m = n - 1$, this variance coincides with $v(\bar{y})$.

In the bootstrap method, the above procedure of selecting the m units is repeated independently a large number (B) of times and the means $\bar{y}_b$, $b = 1, ..., B$, are obtained. Let $\bar{y}_B = \sum_1^B \bar{y}_b / B$ and

$$v_B(\bar{y}) = \sum_1^B (\bar{y}_b - \bar{y})^2 / B. \tag{12.8}$$

This variance is an estimate for $v(\bar{y}_b \mid I)$ and hence for $V(\bar{y})$. For an alternative estimator, $\bar{y}$ in this expression is replaced by $\bar{y}_B$ and the sum of squares is divided by $(B - 1)$.

Finite population

Consider a population of N units with mean $\bar{Y}$ and variance S^2. As seen in Chapter 2, the mean $\bar{y}$ of a sample of n units drawn *without* replacement from this population is unbiased for $\bar{Y}$ and has variance $V(\bar{y}) = (1 - f)S^2/n$, where $f = n/N$. The sample variance s^2 is unbiased for S^2, and an unbiased estimator for $V(\bar{y})$ is $v(\bar{y}) = (1 - f)\,s^2/n$. To estimate this variance, more than one method of adopting the bootstrap can be suggested. They differ in the method of selecting the bootstrap sample and its size. The selection *with* or *without* replacement is examined below.

Selection with replacement

If the bootstrap sample y_i', $i = 1, 2, ..., m$ is selected randomly *with* replacement from the original sample y_i, $i = 1, 2, ..., n$, its mean $\bar{y}_b = \sum_1^m y_i'/m$ is unbiased for $\bar{y}$ and has the variance $v(\bar{y}_b) = (n - 1)\,s^2/nm$. The expression in (12.8) is an approximation for this variance.

If $m = n - 1$, $v(\bar{y}_b)$ is the same as s^2/n. In this case, $v(\bar{y}) = (1 - f)\,s^2/n$ may be estimated from (12.8) by attaching $(1 - f)$ to its right side. On the other hand, $v(\bar{y}_b)$ will be the same as $v(\bar{y})$ if $m = (n - 1)/(1 - f)$, and hence $v(\bar{y})$ can be estimated from (12.8) with this value for m.

Selection without replacement

If the bootstrap sample of size m is selected *without* replacement, $\bar{y}_b$ is unbiased for $\bar{y}$ and $v(\bar{y}_b) = (n - m)s^2/nm$, which will be equal to $v(\bar{y})$ if $m = n/(2 - f)$.

Bootstrap for estimating the variance of the sample ratio

To estimate the variance of the ratio $\hat{R} = \bar{y}/\bar{x}$, the ratios $\hat{R}_b = \bar{y}_b/\bar{x}_b$ are obtained from the bootstrap samples of size m selected from (x_i, y_i), $i = 1, 2, ..., n$. The variance of $\hat{R}$ is obtained from

$$v_B(\hat{R}) = \sum_1^B (\hat{R}_b - \hat{R})^2/B. \tag{12.9}$$

As in the case of the mean, an alternative estimator for the variance is obtained by replacing $\hat{R}$ in this expression with $\hat{R}_B = \Sigma_1^B \hat{R}_b/B$.

To estimate the MSE of $\hat{Y}_R = \hat{R}X$, P.S.R.S Rao and Katzoff (1996) compared the relative merits of the linearization, jackknife, and bootstrap procedures. For the bootstrap procedure, the above approaches and other available methods were considered. In this investigation, all the procedures were found to underestimate the MSE of $\hat{Y}_R$, but the underestimation was the least for the jackknife estimator. Among the bootstrap methods, sampling without replacement with the optimum size $n/(2 - f)$ was found to have relatively the smallest amount of underestimation. The differences in the stabilities of the different estimators were not significant.

12.5 Balanced repeated replication (BRR)

McCarthy (1966, 1969) presents this procedure for estimating the variance of a nonlinear estimator obtained from samples of size two selected *with* replacement from each of G strata. This method for the variance estimator of the population total is examined below.

With $n_g = 2$, an unbiased estimator for the total is obtained from

$$\hat{Y} = \sum_g N_g \bar{y}_g = \sum_g N_g(y_{g1} + y_{g2})/2 = \sum_g (e_{g1} + e_{g2}), \tag{12.10}$$

where $y_{gi}, i = (1, 2)$, are the sample observations and $e_{gi} = N_g y_{gi}/2$. Denoting the variance of y_{gi} by σ_g^2, the variance of this estimator becomes $V(\hat{Y}) = \Sigma N_g^2 \sigma_g^2/2$. Since the sample variance of the gth stratum is given by $s_g^2 = \Sigma_i(y_{gi} - \bar{y}_g)^2 = (y_{g1} - y_{g2})^2/2$, an unbiased estimate of $V(\hat{Y})$ is obtained from

$$v(\hat{Y}) = \sum_g N_g^2(s_g^2/2) = \frac{1}{4}\sum_g N_g^2(y_{g1} - y_{g2})^2 = \sum(e_{g1} - e_{g2})^2.$$

(12.11)

Now, for the BRR method, one observation is selected from the two sample observations of each stratum. From these **half-samples** an estimator for the total is given by $K = \Sigma_g N_g y_{gi}$ where y_{gi} can be the first or the second unit in the sample. Note that $K - \hat{Y} = \Sigma_g \pm (e_{g1} - e_{g2})$. The sign in front of the parenthesis depends on whether the first or the second unit is chosen from the sample of the gth stratum. From this expression,

$$(K - \hat{Y})^2 = v(\hat{Y}) + \text{cross-product terms.} \qquad (12.12)$$

McCarthy (1966) showed that a set of r, $(G + 1 \le r \le G + 4)$, half-samples can be selected such that for every pair of strata half the cross-product terms will have positive signs and the remaining half negative signs. These are known as the **balanced half-samples**. With such samples, from (12.12),

$$\frac{1}{r}\sum_1^r (K_j - \hat{Y})^2 = v(\hat{Y}). \qquad (12.13)$$

With the sample units not included in K, another estimator for the total is given by $L = \Sigma_g N_g y_{gi}$. Using the balanced half-samples from this set, another estimator of variance is obtained by replacing K_j in (12.13) by L_j, $j = (1, ..., r)$. Further, since $\hat{Y} = (K + L)/2$, from (12.13) or otherwise,

$$\frac{1}{4r}\sum_1^r (K_j - L_j)^2 = v(\hat{Y}). \qquad (12.14)$$

Any of these estimates or their averages can be used for estimating $V(\hat{Y})$. However, their biases and stabilities for estimating the MSEs of nonlinear estimators can be different.

Bean (1975) examined the BRR for ratio estimation with two-stage sampling and selection of units with unequal probabilities. Lemeshow and Levy (1978) compared it with the jackknife. Krewski and J.N.K. Rao (1981) compared the BRR with linearization and jackknife for ratio estimation, J.N.K. Rao and Wu (1985) with these methods for stratified sampling, and Valliant (1993) for poststratified estimation. All the three procedures were found to be satisfactory for finding the confidence limits. In some of the empirical studies, linearization seemed to provide slightly better stabilities for the variance estimators than the BRR and jackknife. Wolter (1985) presents this procedure along with the linearization and jackknife methods.

The BRR is closely related to the procedure of estimating the variance through the **interpenetrating subsamples** briefly described in Section 11.1. In this approach, the original sample is divided randomly into k groups of size $m = (n/k)$ each. Denoting their means by $\bar{y}_i$, $i = 1, ..., k$, $\Sigma(\bar{y}_i - \bar{y})^2/k(k - 1)$ is an unbiased estimator for $V(\bar{y})$. Gurney and Jewett (1975) extend the BRR to the case of $n_g > 2$, with equal sample sizes in all the strata.

12.6 Small-area estimation

In several large-scale surveys, estimates for specified areas, regions, or subpopulations are needed. For example, the U.S. Department of Education requires estimates of the total number and proportion of children of 5 to 17 years age from low-income families in each of the school districts in more than 3000 counties. In 1997 to 1998, more than \$7 billion was allocated for the educational programs of children in these types of families. The above estimates are obtained from the CPS of the U.S. Bureau of the Census. Suitable modifications of these estimates are presented in the report edited by Citro et al. (1998).

Sample estimates

In a sample survey conducted in a large area, the sample sizes observed for its small areas can be very small. Chapter 6 examined the estimation for subpopulations. The variance $V(\bar{y}_i) = (1 - f)S_i^2/n_i$ in (6.7) for estimating the mean $\bar{Y}_i$ of the ith subpopulation or area clearly becomes large if n_i is small. Similarly, for small sample sizes, the variances in (6.11) and (6.17) for the estimates of the total and proportion and also the variances of the estimates obtained in Sections 6.7 and 6.8 through stratification can be large.

If the mean of the supplementary characteristic $\bar{X}_i$ for the ith area is known, one can also consider the ratio estimator $(\bar{y}_i / \bar{x}_i)\bar{X}_i$ or the regression estimator $\bar{y}_i + b_i(\bar{X}_i - \bar{x}_i)$ for $\bar{Y}_i$, where $\bar{x}_i$, $\bar{y}_i$, and b_i are obtained from the units observed in that area. However, if the observed sample size n_i is small, these estimators can have large biases and MSEs.

Synthetic and composite estimators

In the synthetic estimation examined by Gonzalez (1973) and others, estimates for a small area are obtained by assuming that it is similar to the entire large area or population. For the mean of the ith area, the ratio estimator $(\bar{y} / \bar{x})\bar{X}_i$ or the regression estimator $\bar{y}_i + b(\bar{X}_i - \bar{x}_i)$, where $(\bar{y}/\bar{x})$ and b are obtained from all the n sample observations, are examples of the synthetic estimator. As can be seen, the biases and MSEs of these estimators will be large if the above assumption is not satisfied. Replacing $(\bar{y}/\bar{x})$ and b by the corresponding quantities obtained from the small areas similar to the ith area can reduce these biases and MSEs.

For composite estimation, a sample estimator is combined with the synthetic estimator. As an example, $W_i\bar{y}_i + (1 - W_i)(\bar{y}/\bar{x})\bar{X}_i$, where the weights W_i are suitably chosen, can be considered for the mean of the ith small area. Similar types of estimators were considered, for example, by Wolter (1979). Schaible et al. (1977) compare the synthetic and composite estimators. Drew et al. (1982) examine the composite estimator with data from the Canadian Labour Force survey. Särndal and Hidiroglou (1989) consider a composite estimator for estimating the wages and salaries for industries in census divisions, and suggest procedures for determining the weights. Longford (1999) presents procedures for multivariate estimation of small-area means and proportions.

Best linear unbiased predictor (BLUP)

For this procedure, the observations of the small areas are considered to constitute a random sample following the model

$$y_{ij} = \mu_i + \varepsilon_{ij} = \mu + \alpha_i + \varepsilon_{ij}, \tag{12.15}$$

where $i = 1, 2, ..., k$ represents the areas and $j = 1, 2, ..., n_i$ represent the observations for the ith area. Note that the mean of the ith area is μ_i, which is expressed as $\mu + \alpha_i$, where μ is the expected value of μ_i and α_i is the **random effect**. The random error ε_{ij} is assumed to have mean zero and variance σ_i^2. The random effect α_i, which is also known as the model error, is assumed to have mean zero and variance σ_α^2 and independent of ε_{ij}.

The BLUP for α_i is considered to be of the form $\hat{\alpha}_i = \Sigma c_{ij} y_{ij} + d$. The coefficients c_{ij} and d are obtained by minimizing $E(\hat{\alpha}_i - \alpha_i)^2$ with the unbiasedness condition $E(\hat{\alpha}_i - \alpha_i) = 0$. The BLUP for α_i can be expressed as $[\sigma_\alpha^2/(\sigma_\alpha^2 + v_i)](\bar{y}_i - \mu)$, where $v_i = V(\bar{y}_i) = \sigma_i^2/n_i$. As a result, the BLUP for μ_i is given by $\hat{\mu}_i = [\sigma_\alpha^2/(\sigma_\alpha^2 + v_i)]\,\bar{y}_i + [v_i/(\sigma_\alpha^2 + v_i)]\mu$. The error of this estimator is given by $E(\hat{\mu}_i - \mu_i)^2 = [\sigma_\alpha^2\,v_i/(\sigma_\alpha^2 + v_i)]$, which is smaller than v_i. For the BLUP, μ can be estimated from the WLS estimator $\Sigma W_i \bar{y}_i / W$, where $W_i = 1/(\sigma_\alpha^2 + v_i)$ and $W = \Sigma W_i$. P.S.R.S. Rao (1997), for instance, presents the derivation of the BLUP and different procedures for estimating σ_α^2 and v_i.

With the assumption that α_i and ε_{ij} follow independent normal distributions, the BLUP becomes the same as the Bayes estimator for the mean of the ith small area derived in Appendix 12A. The Bayes estimator and the BLUP obtained from the sample observations are known as the empirical Bayes estimator (EB) and the empirical best linear predictor (EBLUP). Ghosh and Meeden (1986), for example, describe the EB for the estimation in finite populations. Prasad and J.N.K. Rao (1990), Lahiri and J.N.K. Rao (1995), and Bell (1999) present procedures for estimating the MSE of the BLUP.

Applications, extensions, and evaluations

Estimation procedures for small areas using census data and additional information from administrative records or similar sources were described by Purcell and Kish (1980) and Bell (1996). Ghosh and J.N.K. Rao (1994) review the different procedures suggested for improved estimation in small areas.

In classical statistics, the model in (12.15) is known as the **One-Way Variance Components Model**, and it can be adapted as above to estimate the means and totals of small areas in finite populations. With supplementary variables or predictors x_1, x_2, ..., the mean μ can be replaced by $\beta_0 + \beta_1 x_{1i} + \beta_2 x_{2i} + ...$ Suitable to the application, the intercept β_0 and some of the coefficients β_1, β_2, ..., are assumed to be random. The BLUP can be considered for the resulting model.

To predict the number of children from low-income families in the counties described above, a model of the type

$$y_i = \beta_0 + \beta_1 x_{1i} + \beta_2 x_{2i} + \beta_3 x_{3i} + \beta_4 x_{4i} + \beta_5 x_{5i} + \alpha_i + \varepsilon_i$$

$$(12.16)$$

was employed. In this model, y_i is the logarithm of the 3-year weighted average of the number of children in the ith county. The predictors (x_1, x_2, x_3, x_4, x_5) included the logarithms of the child exemptions reported by families, population under 18 years of age, and other related variables. The sampling error of y_i is represented by ε_i and the model error by α_i. When the ith county is considered to be selected randomly from an infinite population, both these errors become random.

Models of the type in (12.15) and (12.16) were considered by Fay and Herriot (1979) for estimating per capita incomes in regions of the U.S. with fewer than 500 persons, by Ghosh and Lahiri (1987) for small-area estimation with stratified samples, by Dempster and Raghunathan (1987) to estimate wages paid by small business firms, by Battese et al. (1988) for the estimation of agricultural production in segmented areas, and by Stroud (1987) to estimate the average number of times students visit their homes during an academic year. Holt et al. (1979) derive the BLUP for the model of the type in (12.16). Ericksen (1974), Särndal (1984), and Sarndal and Hidiroglou (1985) use the regression-type models for small-area estimation.

The above types of models were also used by Stasny (1991) for a survey on crime, by Stasny et al. (1991) to estimate wheat production in counties, and by Fay and Train (1995) and Fisher and Siegel (1997) for estimating income in different states of the U.S.

Small-area estimation for census undercount was examined by Isaki et al. (1987) and others. Cressie (1989; 1992), Datta et al. (1992), and others consider the EB approach for this purpose. Zaslovsky (1993) uses the EB to estimate the population shares through census data and additional information. Wolter and Causey (1991) examine some of the procedures employed for the estimation of population in small areas. Estimation for small areas from the U.S. National Health Interview Survey was described by Marker (1993) and from the U.S. National Medical Expenditure Survey by Cohen and Braden (1993). Estimation for local areas and regions in Sweden was described by Lundstrom (1987) and Ansen et al. (1988) and in Finland by Lehtonen and Veijanen (1999). Pfeffermann and Burck (1990) and J.N.K. Rao and Yu (1992) present estimation for small areas from surveys conducted over periods of time.

12.7   Complex surveys

Many of large-scale surveys, especially nationwide, employ stratification and clustering of population units, and they are usually conducted in two or more stages. Samples at different stages are selected with equal or unequal probabilities, and the ratio or regression methods are employed for estimation of the population quantities. Nonresponse adjustments of the type described in Chapter 11 are also implemented in several cases. With all these factors, estimation of population quantities, finding standard errors of the estimators, and analysis of the data can become very **complex**.

Regression analysis for complex survey data was described by Holt et al. (1980) and Nathan (1988) among others. J.N.K. Rao and Scott (1981) describe the analysis of categorical data from complex surveys. For qualitative characteristics, Binder (1983) considers logistic regression and estimation of the standard errors. Through an empirical investigation, Katzoff et al. (1989) evaluate some of the procedures for analyzing complex survey data. Some of the methods of analyzing data from complex surveys is presented, for example, in the volume edited by Skinner et al. (1989) and in Lehtonen and Pahkinen (1995). Binder and Patak (1994) present procedures for point and interval estimation with complex survey data. J.N.K. Rao et al. (1992) review the bootstrap, jackknife, and BRR methods for variance estimation and confidence intervals with data from complex surveys. Skinner and J.N.K. Rao (1996) describe the estimation in complex surveys through dual frames. The linearization and jackknife procedures for the inference from complex surveys with multiple frames was described by Lohr and J.N.K. Rao (2000).

Exercises

12.1.   (a) Since the sample mean is unbiased, show that jack-knifing it will again result in the same estimator. (b) For the sample of the 1995 total enrollment in Table T4 in the Appendix, find the ten pseudo values and show that the jackknife estimator coincides with the sample mean.

12.2.   The estimator $(n-1)s^2/n$ is biased for the population variance S^2. (a) Show that jackknifing this estimator results in the unbiased estimator s^2. (b) Verify this result from the sample of the 1995 total enrollments in Table T4 in the Appendix.

12.3. For a sample of ten states in the U.S., the 1995 expenditures for hospital care (y) and physician services (x) are presented in Table T7 in the Appendix. For estimating the ratio of the totals of y and x, compare $\hat{R}$, $\hat{R}_T$, and $\hat{R}_J$.

12.4. For estimation of the ratio in Exercise 12.3, compare the variance estimator in (9.2) and the jackknife estimator in (12.6).

12.5. (a) From the sample of the ten states in Table 12.1, find the regression coefficient b for the regression of the private enrollment (y) on the total enrollment (x). Denote by b_j the regression coefficient obtained by omitting one of the sample observations. The pseudo values are given by $nb - (n - 1)b_j$. The jackknife estimator for the regression coefficient is $b_J = nb - (n - 1)b'$, where $b' = \Sigma b_j/n$. (b) Find this estimate from the above sample, and compare it with the sample regression coefficient. (c) From the pseudo values, find the jackknife variance for b.

12.6. For the problem in Exercise 12.5, generate 100 bootstrap samples of size $m = (n - 1)$ *with* replacement as described in Section 12.4, find the variance estimate of the first type in (12.8), and compare it with the jackknife estimator in Exercise 12.5(c).

12.7. The motivation for formulating the variance estimator in (12.6) is that the pseudo values in (12.3) are approximately uncorrelated. Show that this result is valid.

Appendix A12

The Bayesian approach

Consider a large group or an infinite population following a specified distribution, for example, normal with mean μ_i and variance σ_i^2. For given (μ_i and σ_i^2), the distribution of the mean $\bar{y}_i$ of a sample of size n_i from this population, which can be denoted by $f(\bar{y}_i \mid \mu_i, \sigma_i^2)$, follows the normal distribution with mean m and variance σ_i^2/n_i.

The mean μ_i is assumed to have a *prior distribution* $g(\mu_i)$, for example, normal with mean μ and variance σ_α^2. Now, the *joint distribution* of $\bar{y}_i$ and μ_i are given by $f(\bar{y}_i \mid \mu_i, \sigma_i^2)g(\mu_i)$. The *marginal distribution* of $\bar{y}_i$, denoted by $h(\bar{y}_i)$ is obtained from this joint distribution. The *posterior distribution*

$k(\mu_i \mid \bar{y}_i)$, which is the distribution of μ_i for given $\bar{y}_i$, is obtained by dividing $f(\bar{y}_i \mid \mu_i, \sigma_i^2)g(\mu_i)$ by $h(\bar{y}_i)$.

With the assumption for normality for $f(\bar{y}_i \mid \mu_i, \sigma_i^2)$ and $g(\mu_i)$ as above, $k(\mu_i \mid \bar{y}_i)$ becomes the normal distribution with mean $E(\mu_i \mid \bar{y}_i) = \mu + \beta(\bar{y}_i - \mu)$ and variance $V(\mu_i \mid \bar{y}_i) = \sigma_\alpha^2(\sigma_i^2/n_i)/(\sigma_\alpha^2 + \sigma_i^2/n_i)$. Notice that this mean can also be expressed as $E(\mu_i \mid \bar{y}_i) = a\mu + (1 - a)\bar{y}_i$, where $a = (\sigma_i^2/n_i)/(\sigma_\alpha^2 + \sigma_i^2/n_i)$. The weights for this posterior mean are inversely proportional to σ_α^2 and σ_i^2/n_i, and $V(\mu_i \mid \bar{y}_i)$ is smaller than both these variances. Further, it approaches $\bar{y}_i$ as n_i becomes large. For the EB approach, $E(\mu_i \mid \bar{y}_i)$ is obtained by estimating μ, σ_α^2, and σ_i^2 from the sample observations drawn from all the groups.

In general, $f(\bar{y}_i \mid \mu_i, \sigma_i^2)$ and $g(\mu_i)$ need not be normal. For such cases, $E(\mu_i \mid \bar{y}_i)$ may not be a simple linear combination of the prior mean μ and the sample mean $\bar{y}_i$. For Bayesian analysis, the entire posterior distribution or its mode are examined. For some applications, a prior distribution for σ_i^2 is also specified.

Appendix Tables

Table T1. 1920 random numbers.

756462	782281	651204	875988	737617	420957	200781	690085
730247	841548	773095	289911	695212	731941	656163	238487
175176	423124	584063	480361	678396	442717	789850	935682
547517	862316	794595	654286	579669	392056	516663	556169
216819	615085	770837	320200	932203	172933	846538	672231
354610	402051	341182	433149	215278	522156	966231	367687
579699	483337	510910	276803	582781	321451	738792	768746
539033	834635	566988	464858	267953	702368	735221	752190
684622	798486	859096	155950	395596	331095	440229	355998
516800	294504	794015	652181	587802	125752	832759	140751
960570	351009	538881	180044	319987	757897	705435	893582
735679	303278	355052	532212	915952	306406	104770	224281
572451	300943	144688	437941	723594	324854	920270	785669
358913	463210	236702	474197	396893	592853	419782	855770
330851	508026	505982	155354	359416	779626	165716	379742
715934	862651	531068	775323	857082	624500	107898	697378
723014	943632	640065	616062	177480	390500	311335	713233
754173	194693	548479	350398	349513	919538	701453	443098
264031	582156	349681	312967	303949	715049	770592	310694
786447	655629	259880	996719	812937	750755	982543	453505
680410	864589	363124	264168	165258	365307	999969	268792
668233	110050	893597	524415	180059	423444	921705	909589
866115	678884	171468	148869	268792	714331	914167	243141

430311	838237	263359	549348	118793	548708	266533	964888
696661	813456	603168	441877	753822	317698	663701	746605
902188	383343	816370	374935	711005	574816	889767	888562
552385	981048	424192	654317	271004	136067	216910	910932
116535	761971	482696	806909	767922	556597	945814	225059
966933	517685	928953	807901	270104	845698	686682	215308
113376	810892	708197	581103	702582	101062	899884	212348
892972	356746	221107	954634	308252	162267	609806	756096
205344	911496	342082	593524	486251	916089	630406	315027
559694	485687	177648	500656	795114	369274	686850	927870
817194	396146	263649	381039	419218	368908	480712	487228
431181	707419	307245	156316	616047	203284	882031	188589
791040	894665	682119	483337	651936	234748	555116	952147
800592	207083	835093	251335	499832	387768	484909	927000
558794	890454	723762	261803	327845	307779	709342	961882
813044	387249	719535	366558	958602	148457	308069	918699
917570	200812	913022	187841	979659	134434	826884	403760

Table T2. SAT scores of 30 candidates.

Verbal	Math	Verbal	Math
500	650	390	620
610	630	550	650
560	700	400	650
380	580	640	700
480	590	670	690
460	680	550	750
420	630	530	700
480	630	380	420
510	530	460	590
560	640	560	540
480	670	660	650
500	610	420	450
510	700	470	590
580	630	480	760
540	680	520	750

Table T3. Persons over age 25 in the U.S. (1000s), 1995.

AL	2.8	IN	3.7	MT	0.6	RI	0.6
AK	0.4	IA	1.8	NE	1.0	SC	2.3
AZ	2.0	KS	1.6	NV	1.0	SD	0.4
AR	1.6	KY	2.6	NH	0.7	TN	3.5
CO	2.4	LA	2.7	NJ	5.2	UT	1.1
CT	2.2	ME	0.8	NM	1.0	VT	0.4
DE	0.4	MD	3.3	NC	4.7	VA	4.2
DC	0.4	MA	4.1	ND	0.4	WA	3.4
GA	4.0	MI	6.1	OH	7.2	WV	1.2
HI	0.7	MN	2.9	OK	2.0	WI	3.3
ID	0.7	MS	1.6	OR	2.0	WY	0.3
IL	7.5	MO	3.4	PA	7.9		
CA	19.8	FL	9.6	NY	11.8	TX	11.4

Source: U.S. Bureau of the Census, Current Population Survey, March 1996 and *The New York Times 1998 Almanac*, p. 374, (Ed.) J. W. Wright.

Table T4. Ten states with the largest college enrollments (1000s).

State	1990		1995	
	Total	Public	Total	Public
Illinois	729	551	718	530
Pennsylvania	604	343	618	340
Michigan	570	487	548	462
Ohio	555	426	540	410
Florida	538	440	637	531
Massachusetts	419	185	414	177
Virginia	353	291	356	293
North Carolina	352	285	372	303
New Jersey	324	262	334	271
Wisconsin	300	254	300	246
Mean	474.4	352.4		
S.D.	144.19	117.82		
California	1,770	1,555	1,817	1,564
New York	1,035	608	1,042	588
U.S. 49 states and D.C.	10,858	8,531	14,262	11,092

Source: Statistical Abstract of the United States; 1992, Table 264; 1997, Table 288.

Table T5. Ten states with the smallest college enrollments (1000s).

State	1990		1995	
	Total	Public	Total	Public
Alaska	30	28	29	28
Wyoming	31	31	30	29
South Dakota	34	27	37	30
Montana	36	32	43	37
Vermont	36	21	35	20
North Dakota	38	35	40	37
Delaware	42	34	44	36
Idaho	52	41	60	49
Hawaii	54	43	63	50
Maine	57	42	57	38

Source: Statistical Abstract of the United States; 1992, Table 264; 1997, Table 288.

Table T6. Blood pressures of 20 persons (mmHg).

Before Treatment		After Treatment	
Systolic	Diastolic	Systolic	Diastolic
140	90	130	85
160	100	130	80
145	92	140	85
150	80	150	80
130	92	125	90
180	100	160	85
190	105	160	90
180	105	150	95
170	100	150	85
170	90	160	80
150	85	120	80
140	85	120	80
130	90	130	80
145	85	130	85
145	90	120	80
160	100	140	90
180	100	150	65
180	95	140	80
180	95	130	85
170	90	140	80

Table T7. Physicians in the U.S. in 1995 and hospital expenditure for a sample of ten states.

State	No. of Physicians (1000s)	Expenditures ($10^9) Hospital Care	Physician Services
Arizona	8.3	4.0	2.8
California	75.5	34.8	29.0
Colorado	8.4	3.9	2.5
Florida	31.1	17.1	10.5
Hawaii	2.8	1.5	0.8
Massachusetts	23.5	10.0	4.4
New Hampshire	2.5	1.4	0.8
New Jersey	21.9	10.3	5.8
Pennsylvania	32.9	19.5	7.5
West Virginia	3.6	2.3	1.0

Source: Statistical Abstract of the United States: 1997, Table 179 for the number of physicians and Table 165 for expenditures.

Table T8. Wheat production in 25 countries (million metric tons).

	1990	1995	1997
Burma	0.1	0.2	0.2
Japan	1	0.4	0.6
Bangladesh	1	1.4	1.5
Brazil	3.3	1.5	2.8
South Africa	1.7	2	2.3
Mexico	3.9	3.5	3.8
Hungary	6.2	4.6	5.3
Egypt	4.3	5.7	5.9
Kazakhstan	16.2	6.5	8.7
Italy	8.1	7.7	6.7
Romania	7.3	7.7	7.2
Argentina	11	8.6	14.5
Iran	8	11.3	10
United Kingdom	14	14.3	15.1
Turkey	16	15.5	16
Ukraine	30.4	16.3	18.4
Australia	15.1	16.5	19
Pakistan	14.4	17	16.7
Germany	18	17.8	19.9
Canada	32.1	25	24.3
Russia	49.6	30.1	44.2
France	33.6	30.9	34
United States	74.3	59.4	68.8
India	49.9	65.5	69
China	98.2	102.2	124
Total	517.7	471.6	538.9

Source: Statistical Abstracts of the United States, 1999, Table 1383.

Table T9. Energy production and consumption in 1996 in 25 countries.

| | Production | | Consumption, per capita, mil. BTU |
	Natural Gas, quad. BTU	Petroleum, 1000 barrels per day	
World	81.90	64,054	65
1. United States	19.54	6,465	352
2. Australia	1.12	570	223
3. Austria	0.06	21	158
4. Bahrain	0.26	35	517
5. Belgium	—	—	247
6. Canada	5.97	1,837	407
7. Czech Republic	0.01	4	227
8. Denmark	0.25	208	173
9. Finland	—	—	247
10. France	0.10	43	169
11. Germany	0.70	60	176
12. Japan	0.09	12	170
13. South Korea	—	—	157
14. Kuwait	0.22	2,062	319
15. Netherlands	3.01	56	242
16. New Zealand	0.20	37	247
17. Norway	1.61	3,104	397
18. Russia	19.35	5,850	176
19. Saudi Arabia	1.53	8,218	214
20. Sweden	—	—	255
21. Switzerland	—	—	172
22. Trinidad and Tobago	0.32	—	284
23. United Arab Emirates	1.34	2,278	793
24. United Kingdom	3.34	2,568	171
25. Taiwan	0.04	1	145

Source: Statistical Abstract of the United States, 1998, Table 1379.

Solutions to Exercises

Chapter 1

1.1 If a public gathering can be considered to be a representative sample of the entire population, estimates derived from the show of hands or applause can provide valid estimates.

1.3 (1) The people contacted should be representative of the population.

(2) Persons should be contacted during the peak periods and also other times during the week.

1.5 Proportional distribution of the sample to the four classes provides better estimates; see Chapter 5.

1.7 Systematic sampling should be recommended for the characteristics with an approximate linear trend. Otherwise, both systematic and simple random sampling provide almost the same precision.

1.9 (1) The purchases for these items should be large in number.

(2) The returned questionnaires should be representative of all the consumers who purchased the products.

1.11 (1) To estimate the percentage of the purchases, add the numbers of purchases in the two groups, subtract the number of purchases from the 30 duplicated units from the total, and divide the resulting number by 270.

(2) Similarly, to estimate the average of the purchases, add the totals for the two groups, subtract the amount for the 30 duplicated units, and divide the resulting amount by 270.

Chapter 2

2.1 (a) For the 80 corporations, $\hat{Y} = 20,000$.

(b) S.E. ($\hat{Y} = 1,704$).

(c) 95% confidence limits are (16660, 23340).

2.3 (a) S.E. of the mean is $15/(30)^{1/2} = 2.74$, $t_{29}(.05) = 2.0452$, and 95% confidence limits are (29.4, 40.6).

(b) C.V. of the sample mean is equal to $2.74/35 = 0.09$, that is, 9%.

2.5 (a) $V(t_1) = [Nn_2/(N_1 + n_2)]^2 V(\bar{y}_2)$. From the sample, $v(\bar{y}_2) = 20$, $v(t_1) = 20,663.7$ and S.E.$(t_1) = 143.75$.

(b) $E(t_1) = N(Y_1 + n_2\bar{Y}_2)/(N_1 + n_2)$, which is not equal to $Y = Y_1 + Y_2$, and hence t_1 is biased for Y.

2.7 (a) $n = 32$

(b) $n = 43$

(c) $n = 318$.

2.9 For the three types of households, the required sample sizes are (1) $n = 136$, (2) $n = 169$, and (3) $n = 211$.

2.11 The result in (c) follows from noting that a pair of units appear in $_{N-2}C_{n-2}$ of the $_NC_n$ samples.

Chapter 3

3.1 For the sample sizes 150 and 300, the variances are $V_1 = (3000 - 150)S^2/(3000 \times 50) = 63.33S^2/10^4$ and $V_2 = (3000 - 300)S^2/(3000 \times 300) = 30S^2/10^4$.

(a) Relative decrease in the variance is $(V_1 - V_2)/V_1$, which is 53%.

(b) The S.E.s respectively are equal to $7.96S/100$ and $5.48S/100$, and hence the relative decrease in the S.E. is equal to $(7.96 - 5.48)/7.96 = 0.31$, that is, 31%.

(c) 31%, as in (b).

(d) Relative increase in the precision is $(V_1/V_2) - 1 = 1.11$, that is, 111%.

(e) Confidence widths for the two cases are $2Z(7.96S)/100$ and $2Z(5.48S)/100$. Relative decrease in the confidence width is equal to 0.31, that is, 31%.

(f) Same as (a)–(e) for the estimation of the total.

3.3 (a) $d = \bar{Y}_1 - \bar{y}_2 = 50$.

(b) $V(d) = V(\bar{y}_2) = 453.75$ and hence S.E.$(d) = 21.3$.

(c) The 95% confidence limits for $\bar{Y}_1 - \bar{Y}_2$ are (8.25, 91.75).

(d) Since $E(\bar{Y}_1 - \bar{y}_2) = \bar{Y}_1 - \bar{Y}_2$, it is unbiased.

3.5 (a) 20.

(b) 30.91.

(c) -40.58, 80.58

3.7 First list: simple random sampling can be recommended for all the three items.

Second list: systematic sampling for (1) and (2), and simple random sampling for the third list can be recommended.

Third list: same as for the second list.

3.9 (1) Bias, variance, and MSE of $\bar{y}_1$ are given by $B(\bar{y}_1) = \bar{Y}_1 - \bar{Y} = W_2(\bar{Y}_1 - \bar{Y}_2)$, $V(\bar{y}_1) = (N_1 - n_1)S_1^2/N_1 n_1$ and $MSE(\bar{y}_1) = V(\bar{y}_1) + B^2(\bar{y}_1)$. The bias will be small if $\bar{Y}_2$ is close to $\bar{Y}_1$, and the S.E. will be small if n_1 is large.

(2) Same as (1).

(3) Since the expected value of the resulting estimator is $W_1\bar{Y}_1$, it has a bias of $W_1\bar{Y}_1 - \bar{Y} = -W_2\bar{Y}_2$, which will be small if the size of the nonrespondents is not large. This estimator has a variance of $(n_1/n)^2(N_1 - n_1)S_1^2/N_1 n_1$, which will be small if n is large.

3.11 (a) For the mean of the systolic pressure, 90% confidence limits are (152.8, 166.7).

(b) The hypothesis that the mean is greater than 165 is not rejected at the 5% level of significance.

3.13 The population C.V. of y is $S_y/\bar{Y} = 60.66/670 = 0.09$, that is 9%. For the sample C.V., $s_y/\bar{y}$, the expectation, bias, variance, and MSE, respectively, are equal to 0.08, 0.01, 0.0018, and 0.0019.

The population correlation of x and y is $\rho = 0.41$. For the sample correlation coefficient, the expectation, bias, variance, and MSE are equal to 0.39, 0.02, 0.3853, and 0.3857.

3.15 (a) From the Cauchy-Schwartz inequality, $S_{xy}^2 \leq S_x^2 S_y^2$, and hence $\rho^2 \leq 1$, that is, $-1 \leq \rho \leq 1$.

(b) Similarly, $s_{xy}^2 \leq s_x^2 s_y^2$, and hence $r^2 \leq 1$, that is, $-1 \leq r \leq 1$.

3.17 (c) The unbiasedness follows from (b) by noting that the probability of selecting a pair of units is $n(n - 1)/N(N - 1)$.

3.19 (a) Since each of the population units is selected with equal probability $p_i = 1/N$, $E(y_i) = \sum_1^N p_i y_i = \sum_1^N y_i/N = \bar{Y}$.

(b) $V(y_i) = E(y_i - \bar{Y})^2 = \sum_1^N p_i(y_i - \bar{Y})^2 = (N - 1)S^2/N = \sigma^2$.

(c) Since the units are replaced, they are selected indepen-
dently. Hence, $\text{Cov}(y_i, y_j) = E(y_i - \bar{Y})(y_j - \bar{Y}) = E(y_i - \bar{Y}) \times E(y_j - \bar{Y}) = 0$. This result can also be obtained by noting
that $E(y_i y_j) = E(y_i)E(y_j) = \bar{Y}^2$.

Chapter 4

4.1 (a) Public institutions: $p = 6/10$, S.E.$(p) = 0.1456$, $\hat{C} = 30$,
and S.E.$(\hat{C}) = 7$.
Private institutions: $p = 1/10$, S.E.$(p) = 0.09$, $\hat{C} = 5$, and
S.E.$(\hat{C}) = 4$.
(b) Estimate of the difference in the proportions for the
two types of institutions is $d = 0.5$ and S.E. $(d) = 0.145$.
   Estimate for the difference in the numbers of insti-
tutions with enrollments exceeding 100,000 is $49(.5) = 24.5$, that is, 25, and this estimate has a S.E. of $49(.145) = 7$.

4.3 For the first poll, $e^2 = (1.96)^2(.49)(.51)/597 = 0.0016$.
Hence $e = 0.04$, that is, 4%.
   For the second poll, $e^2 = (1.96)^2(.44)(.56)/597 = 0.0016$, and hence $e = 0.04$ again.
   For the last two polls, $e^2 = (1.96)^2(.57)(.43)/500 = 0.0019$. Hence $e = 0.044$, that is, close to 4.5%.

4.5 (a) $p = 0.31 + 0.12 = .43$, that is, 43%.
(b) $v(p) = (.43)(.57)/499 = 0.0005$. Hence S.E.$(p) = 0.022$,
that is 2.2%.
(c) The 95% confidence limits are $(39, 47)$%.
(d) Denoting the proportions in favor and not in favor by
P and $Q = 1 - P$, their difference is $D = 2P - 1$. The null
and alternative hypotheses are $D \leq .10$ and $D > .10$.
   From the sample results, an unbiased estimate of this
difference is $2p - 1 = .14$, which has the sample variance
$v(2p - 1) = 4v(p) = 4(.57)(.43)/499 = 0.002$ and hence the
S.E. of 0.045.
   Now, $Z = (.14 - .10)/.045 = 0.89$, which is less than
1.65, the percentile of the standard normal distribution
for the one-sided 5% level of significance. Hence, we do
not reject the null hypothesis that $(P - Q)$ is less than 10%.

4.7 (a) For the Democrats, $p = .21 + .22 + .31 = .74$, that is,
74%. Since $v(p) = (.74)(.26)/1697 = .00011$, S.E. $(p) = .01$,
that is, 1%. The 95% confidence limits are $(72, 76)$%.

(b) For the Republicans, $p = 70\%$, S.E.$(p) = 1.2\%$, and 95% confidence limits are (67.6, 72.4).

(c) For the Democrats and Republicans together, $p = [1698(.74) + 1432(.70)/3130 = .722$, that is, 72.2%. Now, $v(p) = (.722)(.278)/3129 = 64 \times 10^{-6}$, and hence S.E.$(p) = 0.8\%$. 95% confidence limits for this percentage are (70.6, 73.8)%.

d) An estimate for the percentage without college education is $100 - 72.2 = 27.8\%$. It has the same S.E. of 0.8 as in (c). Confidence limits for this percentage are (26.2, 29.4).

4.9 (a) For the Freshman–Sophomore, Junior, and Senior groups, the error e is $.2/5 = .04$, $.5/5 = .1$, and $.8/5 = .16$, respectively.

The required sample sizes are 322, 88, and 24, respectively. For the first group, the small value of .04 for the error resulted in a relatively large sample size.

(b) With $e = .10$, same for each group, the required sample sizes for the three groups are 60, 88, and 59. Since $p = .5$ for the Juniors, sample size required for this group is relatively larger.

4.11 (a) 82.

(b) 197.

(c) 188.

4.13 Preliminary estimates of the proportions in favor of continuing the program are $p_1 = 8/11$, $p_2 = 12/25$, and $p_3 = 1/2$.

(a) $V(p_1 - p_2) = V(p_1) + V(p_2)$. Replacing the observed sample sizes n_1 and n_2 in this expression by their expected values nW_1 and nW_2, where $W_1 = N_1/N$ and $W_2 = N_2/N$, this variance approximately becomes equal to $[(N - n)/Nn][P_1Q_1/W_1 + P_2Q_2/W_2]$. Since this variance should not exceed $(.07)^2$, $n = 698$.

(b) As in (a), sample sizes required to estimate $(P_1 - P_3)$ and $(P_2 - P_3)$ with the S.E.s of the estimates not exceeding 0.07, the required sample sizes are 1001 and 280.

A sample of size 1001 from the 8500 units is required to estimate the three differences of the percentages with the S.E. not exceeding 7% in each case.

4.15 (a) From (4.20), $E(c_1c_2) = n(n - 1)C_1C_2/N(N - 1)$. Now, $\mathrm{Cov}(c_1, c_2) = E(c_1c_2) - n^2 P_1P_2 = -[(N - n)n/(N - 1)]P_1P_2$. Hence, $\mathrm{Cov}(p_1, p_2) = -[(N - n)/(N - 1)n]P_1P_2$.

(b) An unbiased estimator for P_1P_2 is given by $(N - 1) \times np_1p_2/N(n - 1)$. Substituting this expression in the above covariance, its unbiased estimator is given by $-[(N - n)/N(n - 1)]p_1p_2$.

Chapter 5

5.1 Difference of the means: (a) 10.32.
(b) 3.81.
(c) (2.92, 17.86).
Difference of the totals: (a) 99.03.
(b) 42.88.
(c) (14.99, 183.1).

5.3 For a simple random sample of size 16 from the 43 units, $V(\bar{y}) = 1.243$.
For proportional allocation, $n_1 = 12$ and $n_2 = 4$, $V_P = 0.393$, and the relative precision is $1.243/.393 = 3.16$.
For Neyman allocation, $n_1 = 5$ and $n_2 = 11$. We select five units from the first stratum and *all* the 11 units of the second stratum. This is an example of 100% sampling from a stratum.
From (5.11), the estimator for the population mean now can be expressed as $\hat{\bar{Y}}_{st} = W_1\bar{y}_1 + W_2\bar{Y}_2$. Thus, $V_N = V(\hat{\bar{Y}}_{st}) = W_1^2 V(\bar{y}_1) = 0.074$, and the precision of Neyman allocation relative to simple random sampling is $1.243/.074 = 16.80$.

5.5 In Table T2, math scores are below 600 for the eight units (4, 5, 9, 23, 24, 25, 26, 27) and 600 or more for the remaining 22 units. For the total score, the variances of the first and second stratum, respectively, are 11541.07 and 8872.73.
When $n = 10$, proportional allocation results in $n_1 = 3$ and $n_2 = 7$. With these sample sizes, from (5.12), $V(\hat{\bar{Y}}_{st}) = 635.74$. Neyman allocation results in the same sample sizes for the strata, and hence $V(\hat{\bar{Y}}_{st}) = 635.74$.
From (5.23) and (5.26), $V_P = 638.95$, and $V_N = 635.60$. The sample sizes are rounded off for finding the actual variance $V(\hat{\bar{Y}}_{st})$ from (5.12). Since the expressions for V_P and V_N ignore the rounding off, their values are slightly different from 635.74.

5.7 Since $W_1 = 2/10$, $W_2 = 8/10$, $W_1 S_1^2 + W_2 S_2^2 = 3060$. As described in Section 5.13, $n = 30$. Proportional allocation results in $n_1 = 6$ and $n_2 = 24$.

5.9 Proportional allocation: $n = 30$. $n_1 = 5$, $n_2 = 7$, and $n_3 = 18$.
Neyman allocation: $n = 29$. $n_1 = 3$, $n_2 = 5$, and $n_3 = 21$.

5.11 $n = 97$. $n_1 = 9$, $n_2 = 18$, and $n_3 = 70$.
$V_{opt} = 0.01424$ and S.E.$(\hat{\bar{Y}}_{st}) = 0.12$.

5.15 (a) Denote the probability of observing the n_g units by P_g.
With $a_g = (P_g/n_g)^{1/2}$ and $b_g = (P_g n_g)^{1/2}$, from the Cauchy-Schwartz inequality, $(\Sigma P_g/n_g)(\Sigma P_g n_g) \geq (\Sigma p_g)^2 = 1$. This final result is the same as $E(1/n_g) \geq 1/E(n_g)$.
(b) Since the covariance of n_g and $1/n_g$ is negative, $E(n_g) \times (1/n_g) - E(n_g)E(1/n_g) \leq 0$. Hence $1 - E(n_g)E(1/n_g) \leq 0$, that is, $E(1/n_g) \geq 1/E(n_g)$.
(c) This result follows since $1/n_g$ is replaced by $1/E(n_g)$ for the average variance.

Chapter 6

6.1 (a) $N_2 = 67$, $\hat{Y}_2 = 25199$, and S.E.$(\hat{Y}_2) = 3914$.
(b) $\hat{N}_2 = 60$, $\hat{Y}_2 = 22566$, and S.E.$(\hat{Y}_2) = 3464$.

6.3 (a) and (b)

Estimates and S.E.s

	Sizes	Proportion	Total
Owners	Known		
	8549	.20, .0254	1710, 217
	Estimated		
	8897	.20, .0254	1779, 226
Renters	Known		
	6157	.158, .0287	973, 177
	Estimated		
	5809	.158, .0287	918, 167
Diff.	Known	.04, .0385	737, 280
	Estimated	.04, .0384	861, 281

(c) $d = (58 - 42)/158 = 0.0253$, and S.E.$(d) = 0.0124$.

6.5 $\bar{y} = \Sigma f_i y_i / n = 962/400 = 2.4$
$s^2 = \Sigma f_i (y_i - \bar{y})^2 / (n-1) = 1.5$
$v(\bar{y}) = [(14706 - 400)/14706 \times 400](1.5) = 0.0036$, and hence S.E.$(\bar{y}) = 0.06$.

6.7 For the males, the sample mean is $\bar{y}_M = [5(3.4) + 3(4.67) + 4(4)] = 3.92$. With the expression similar to (5.6), the sample variance for the males is $s_M^2 = 1.69$. Now, $v(\bar{y}_M) = 0.14$ and hence S.E.$(\bar{y}_M) = 0.3742$.

Similarly, for the females, $\bar{y}_F = 2$, $s_F^2 = 0.64$, $v(\bar{y}_F) = 0.08$, and S.E.$(\bar{y}_F) = 0.28$.

Now, $\bar{y}_M - \bar{y}_F = 1.92$, $v(\bar{y}_M - \bar{y}_F) = 0.14 + 0.08 = 0.22$, and hence S.E.$(\bar{y}_M - \bar{y}_F) = 0.47$.

6.9 (a) For the women, the estimate is 43% with a S.E. of 1.6%. For the men, the estimate is 9% with a S.E. of 1.0%.

(b) Estimate for the difference of the percentages of the women and men is 34% with a S.E. of 1.8%.

6.11 (a) The responses now refer to the 850 women and 850 men. In this case, variance of the percentage for the women is $(9.43)(.57)/849 = 288 \times 10^{-6}$, and hence it has a S.E. of 1.7%.

Variance of the percentage for the men is 98×10^{-6} and hence its S.E. is close to 1.0%.

(b) The variance of the difference of the percentages is 384×10^{-6} and hence it has a S.E. of 1.96%.

The estimates and the S.E.s in this case do not differ much from those in Exercise 6.9, since the sample size is large and a large number, 85%, of the sampled units responded.

Chapter 7

7.1 For the establishments, $\bar{\bar{Y}} = 3.02$, $S_b^2 = 6(21.0493)/8 = 15.79$.

From (7.11), $V(\bar{\bar{y}}) = [(9-3)/27](15.79/6) = 0.5848$ and hence S.E.$(\bar{\bar{y}}) = 0.76$.

For a simple random sample of 18 counties, $V(\bar{\bar{y}}) = [(54-18)/54 \times 8](15.94) = 0.5904$ and hence S.E.$(\bar{\bar{y}}) = 0.77$.

Both cluster sampling and simple random sampling have almost the same precision.

7.3 *Employment*:

For the sample clusters (1, 7, 9), from Table 7.1, $\bar{\bar{y}} = (74.2 + 38.5 + 76.3)/3 = 63$ and $s_b^2 = 2707.4$. From (7.15), $v(\bar{\bar{y}}) = (6/27)(2707.7416 = 100.29$.

From Table 7.2, $s_W^2 = (17587.70 + 2378.86 + 2876.83)/3 = 7614.46$. Now, from Appendix A7.2, $s^2 = (15/17)(7614.46) + (2/17)(2707.74) = 7037.20$. For a simple random sample of 18 clusters, $v(\bar{\bar{y}}) = [(54-18)/54 \times 18](7037.20) = 260.64$.

Precision of cluster sampling relative to simple random sampling is $260.64/100.29 = 2.6$.

Establishments:

For cluster sampling, $\bar{\bar{y}} = 4.49$, $s_b^2 = 20.73$, and $v(\bar{\bar{y}}) = 0.77$.

Further, $s_w^2 = 30.03$, $s^2 = 28.93$, and for simple random sampling, $v(\bar{\bar{y}}) = 1.07$. Relative precision of cluster sampling is $1.07/.77 = 1.39$.

7.5 Forty-six of the 54, that is, 85% of the clusters have more than 10,000 employees.

For the three sample clusters (2, 3, 8), $M_i = (8, 3, 7)$, $c_i = (6, 2, 5)$, and $p_i = (.75, .67, .71)$. Hence $\bar{c} = 13/3$ and $p = 0.71$.

The variance of the c_i is $s^2 = [(6 - 13/3)^2 + (2 - 13/3)^2 + (5 - 13/3)^2]/2 = 13/3$. Now, $v(\bar{c}) = [(9 - 3)/27] \times (13/3) = 26/27$, $v(p) = v(\bar{c})/36 = 0.0267$, and hence S.E.$(p) = 0.1636$.

For the ratio method, from (7.27), $\hat{P} = 13/18 = 0.7222$. From (7.29), $v(\hat{P}) = 250 \times 10^{-6}$ and hence S.E.$(\hat{P}) = 0.0158$.

7.9 With the subscript $i = 1, 2, 3, 4$ representing the four clusters, $Q_i = 0.419, 0.503, 0.503$, and 0.575, respectively.

Q_{ij} for $(i, j) = (1, 2), (1, 3), (1, 4), (2, 3), (2, 4)$, and $(3, 4)$, respectively are $0.1292, 0.1292, 0.1607, 0.1667, 0.2071$, and 0.2071.

7.11 From (7.37), $v_1(\hat{Y}_{HT}) = 3134.75$ and S.E.$(\hat{Y}_{HT}) = 55.99$. From (7.38), $v_2(\hat{Y}_{HT}) = 4695.22$ and S.E.$(\hat{Y}_{HT}) = 68.52$.

Since $M_0 = 21$, S.E.s of $\bar{\bar{Y}}_{HT}$ from these two procedures are $55.99/21 = 2.67$ and $68.52/21 = 3.26$, respectively.

Chapter 8

8.1 For the employments, $S_1^2 = 620.8179$ and $S_2^2 = 5294.77$. Since $N = 9$, $n = 3$, and $m = 3$, from (8.4), $v(\bar{\bar{y}}) = (6/27)S_1^2 + (3/54)S_2^2 = 432.11$.

Since $S^2 = 5057.79$, for the mean of a simple random sample of nine counties, $v(\bar{y}) = [(54 - 9)/54 \times 9](5057.79) = 468.31$.

Relative precision of cluster sampling is $468.31/432.11 = 1.084$.

8.3 The first and second terms of (8.9) are 1,266,722.61 and
1,671,459.15. Thus, $V(\hat{\bar{Y}})$ = 2,938,181.76 and S.E. $\hat{Y}$ =
1714.11. Hence, S.E. $\bar{\bar{Y}}$ = 1714.11/54 = 31.74.

8.5 (a) The first term of (8.12) is equal to 510,148.15 and the
second term is the same as in Exercise 8.3. Thus,
$V(\hat{Y}_R)$ = 2,1181, 607.3 and S.E. $(\hat{Y}_R)$ = 1,477.03. Hence,
S.E.$(\bar{\bar{Y}})$ = 1477.03/54 = 27.35.
(b) The above S.E.s for the ratio estimation are about 86%
of the corresponding S.E.s in Exercise 8.3.

8.7 From (8.17), $\hat{C}$ = (9/3)[10/3 + 4/2 + 7(1)] + 37. The first
and second terms of (8.18) are equal to 181.667 and 10.778.
Thus, $v(\hat{C})$ = 192.45 and S.E.$(\hat{C})$ = 13.87.
     The estimate of the proportion is $\hat{P}$ = 37/54 = 0.69
and S.E.$(\hat{P})$ = 13.87/(54) = 0.26.
     Since $\bar{p}$ = 37/66 = 0.56, from (8.19), $\hat{C}_R$ = 54(37/66) =
30.27. The first term of (8.20) is equal to 140.093 and the
second term is equal to 10.778, as above. Thus, $v(\hat{C}_R)$ =
150.871 and S.E.$(\hat{C}_R)$ = 12.28. Hence, S.E.$(\bar{p})$ = 12.28/54 =
0.2274.
     The S.E.s of the estimates of the total and proportion
with the adjustment for the cluster sizes are about 87%
of the S.E.s without the adjustment.

Chapter 9

9.1 (a) R = 670/550 = 1.2182. $E(\hat{R})$ = 1.2211, and hence $\hat{R}$ has
a bias of 0.0029.
(b) $V(\hat{R})$ = 0.004274 and MSE$(\hat{R})$ = 0.004282. From the
approximation in (9.1), $V(\hat{R})$ = 0.004225.
(c) $E[v(\hat{R})]$ = 0.004136. The bias of $v(\hat{R})$ for estimating
$V(\hat{R})$ in (9.1) and the exact MSE $(\hat{R})$ are respectively equal
to -89×10^{-6} and -46×10^{-6}.

9.3 (a) R = 670/(670 + 550) = 0.5492. $E(\hat{R})$ = 0.5547, and
hence $\hat{R}$ has a bias of 0.0055.
(b) The exact variance and MSE of $\hat{R}$ are equal to 140 ×
10^{-6} and 170 × 10^{-6}. With an expression similar to (9.1),
approximations to the variance and S.E. of $\hat{R}$ are equal
to 87 × 10^{-6} and 0.0093, respectively.

9.5 Denoting heights and weights by (x_i, y_i), $R = \bar{Y}/(\bar{X} + \bar{Y})$.
For the five sample units, $(\bar{x}, \bar{y})$ = (66, 158.6) and $\hat{R}$ =
158.6/224.6 = 0.71.

With $t_i = x_i + y_i$, the sample variance of t_i is $s_t^2 = s_y^2 + s_x^2 + 2s_{xy} = 314.8 + 24 + 2(84.5) = 507.8$, and $s_{yt} = s_{yx} + s_y^2 = 84.5 + 314.8 = 399.3$.

Now, $s_d^2 = s_y^2 + \hat{R}^2 s_t^2 - 2\hat{R}s_{yt} = 3.78$. Following (9.2), $v(\hat{R}) = (1/224.6)^2(10/75)(3.78) = 9.99 \times 10^{-6}$, and hence S.E.$(\hat{R}) = 0.0032$.

The 95% confidence limits for R are given by (.70, .72).

Physicians consider the ideal weight to be 100 pounds for a person five feet tall, and five pounds more for each additional inch of height. For healthy persons over five feet, the above ratio should not be much smaller than one.

9.7 Denoting weight, systolic and diastolic pressures by (x, y_1, y_2), the sample means are $(\bar{x}, \bar{y}_1, \bar{y}_2) = (149, 138, 88.6)$.

Now, $\hat{R}_1 = 138/149 = 0.9262$, $\hat{R}_2 = 88.6/149 = 0.5946$, and $\hat{R}_1 - \hat{R}_2 = 0.3315$. From the sample observations, $s_d^2 = 39.9149$. From (9.7), $v(\hat{R}_1 - \hat{R}_2) = (2/15)(39.9149)/(149)^2 = 0.00024$, and hence S.E.$(\hat{R}_1 - \hat{R}_2) = 0.0155$. The 95% confidence limits for $(R_1 - R_2)$ are given by (.30, .36).

9.9 With (x, y) denoting the totals for the number of establishments and the employment, $\bar{X} = 18.1$, $\bar{Y} = 265.31$, and $R = 14.66$. Further, $S_x^2 = 94.72$, $S_y^2 = 21396.18$, and $\rho = 0.9429$.

From the above values, for the ratio estimator of the average employment per cluster, $V(\hat{\bar{Y}}_R) = (1 - f) \times (2395.99)/n$. Since $M = 6$, $V(\bar{Y}_R) = V(\hat{\bar{Y}}_R)/36$.

For a simple random sample of n clusters, $V(\bar{y}) = (1 - f) \times (21396.18)/n$ and $V(\bar{\bar{y}}) = V(\bar{y})/36$.

Hence, precision of the ratio estimator relative to the sample mean is equal to $21396.18/2395.99 = 8.93$.

9.11 Denote the public, private, and total enrollments by (x_1, y_1, t_1) for 1990 and by (x_2, y_2, t_2) for 1995.

(a):

	x_1	y_1	x_2	y_2
Mean	352.4	122	356.3	127.4
S.D.	117.82	76.85	121.77	80.03

Mean and S.D. of t_1 are (474.4, 144.19) and of t_2, they are (483.7, 146.66).

Correlations of the public and total enrollments for 1990 and 1995, respectively, are 0.847 0.838. Further, $\text{Cov}(x_1, t_1) = V(x_1) + \text{Cov}(x_1, y_1) = 1483.49$.

Covariances

	x_1	y_1	x_2	y_2
x_1	13,881.0			
y_1	502.6	5,905.3		
x_2	13,799.1	−36.6	14,828.0	
y_2	661.3	6,140.8	136.0	6404.0

Correlations

	x_1	y_1	x_2	y_2
x_1	1			
y_1	.056	1		
x_2	.962	−.004	1	
y_2	.070	.999	.014	1

(b) and (c):
For 1990, $R = 352.4/474.4 = 0.7428$. The ratio estimator
for the public enrollments is obtained from $\bar{X}_{1R} = (\bar{x}_1/\bar{t}_1) \times (474.4)$. Further, $V(\hat{\bar{X}}_{1R}) = 597.67$ and S.E.$\bar{X}_{1R} = 24.45$.

For the sample mean, $V(\bar{x}_1) = 2082.14$, and hence
S.E.$(\bar{x}_1) = 45.63$. Relative precision of $\hat{\bar{X}}_{1R}$ is $2082.14/597.67 = 3.48$.

For 1995, $R = 356.3/483.7 = 0.7366$. The ratio estima-
tor for the public enrollments is obtained from $\hat{\bar{X}}_{2R} = (\bar{x}_2/\bar{t}_2)(483.7)$. Further, $V(\hat{\bar{X}}_{2R}) = 667.67$ and S.E.$(\hat{\bar{X}}_{2R}) = 25.84$.

For the sample mean, $V(\bar{x}_2) = 2224.20$, and hence
S.E.$(\bar{x}_2) = 47.16$. Relative precision of $\hat{\bar{X}}_{1R}$ is $2224.20/667.67 = 3.33$.

9.13

(a):

	x_1	y_1	x_2	y_2
Mean	33.4	7.6	35.4	8.4
S.D.	7.14	5.34	9.26	6.06

Covariances

	x_1	y_1	x_2	y_2
x_1	50.93			
y_1	8.96	28.49		
x_2	60.82	9.07	85.82	
y_2	14.93	31.51	13.93	50.64

Correlations

	x_1	y_1	x_2	y_2
x_1	1			
y_1	.24	1		
x_2	.92	.76	1	
y_2	.35	.78	.25	1

(b) For 1990, R_1 = 33.4/7.6 = 4.39, and $V(\hat{R}_1)$= 1.35. For 1995, R_2 = 35.4/8.4 = 4.21 and $V(\hat{R}_2)$ = 1.61.

(c) $V(\hat{R}_1 - \hat{R}_2)$ = 0.42.

9.17 Regression of systolic pressure on weight

Variance of ε_i

	Constant	Prop. to x_i	Prop. to x_i^2
b	0.9082	0.9087	0.9091
Reg. SS	301487.9	1926.5	12.4
$\hat{\sigma}^2$	25.9	0.17	0.00112

For each case, $F_{1,14}$ = Reg.MS/$\hat{\sigma}^2$ is very large, indicating that the regression is highly significant, that is, the hypothesis that β = 0 can be easily rejected.

Chapter 10

10.1 After performing (1) through (4), we found that the regression coefficient and the regression estimator for this data are badly biased and have unacceptably large variances and MSEs.

For the data in Table 2.1, we find that β = 0.3254 and ρ = 0.41. Further, the regression and residual MSEs are 3827.04 and 3091.85 with 1 and 4 d.f. Thus $F_{1,4}$ = 3827.04/3091.85 = 0.81, which is not significant even at a very large value for the significance level.

Regression through the origin with $V(y_i|x_i)$ to be constant or proportional to x_i and the corresponding estimator for the mean can be suitable for this data; see Exercises 9.1 and 9.2.

10.3 (a) The regression estimators for the means of the systolic and diastolic pressures are $\hat{\bar{Y}}_{1l}$ = 132.88 and $\hat{\bar{Y}}_{2l}$ = 85.72. Thus, $\hat{\bar{Y}}_{1l} - \hat{\bar{Y}}_{2l}$ = 47.16.

(b) $V(\hat{\bar{Y}}_{1l} - \hat{\bar{Y}}_{2l}) = 11.13$ and S.E.$(\hat{\bar{Y}}_{1l} - \hat{\bar{Y}}_{2l}) = 3.34$.

(c) The 95% confidence limits for the difference of the means are (40.61, 53.71).

10.5 (a) For 1990, $V(\hat{\bar{Y}}_l) = 588.39$ and S.E.$(\hat{\bar{Y}}_l) = 24.26$. For 1995, $V(\hat{\bar{Y}}_l) = 662.27$ and S.E.$(\hat{\bar{Y}}_l) = 25.73$.

(b) For both the years, the variances for the regression estimators do not differ much from 597.67 and 667.67 for the ratio estimators found in Exercise 9.11.

10.7 Denoting the public, private, and total enrollments by (y_1, y_2, x), for large n, the difference of the regression estimators for the public and private enrollments can be approximately expressed as $(\bar{Y}_{1l} - \bar{Y}_{2l}) = (\bar{y}_1 - \bar{y}_2) + (\beta_1 - \beta_2)(\bar{X} - \bar{x})$. Thus, $V(\hat{\bar{Y}}_{1l} - \hat{\bar{Y}}_{2l}) = [(1 - f)/n](s_1^2 + s_2^2 + 2s_{12}) + (\beta_1 - \beta_2)^2 s_0^2 + 2(\beta_1 - \beta_2)(s_{10} - s_{20}) = 11.79$. Hence, S.E.$(\bar{Y}_{1l} - \bar{Y}_{2l}) = 3.43$.

Chapter 11

11.1 $p_1 = 150/250 = 0.6$, $v(p_1) = (.6)(.4)/249 = 9.64 \times 10^{-4}$ and hence S.E.$(p_1) = .031$. Thus $p_1 = 60\%$, and it has a S.E. of 3.1%.

The 95% confidence limits are (54, 66)%.

11.3 Denoting husband's and wife's age by h and w, and income by y, we find the following results from the eight completed observations:

Regression with the intercept:

$\hat{y} = -12.35 + 0.45\ w + 1.30\ h$.

The S.E. for the intercept and the two slopes are 25.6, 2.75 and 2.95.

The total, regression and residual SS are 1737.8, 920.2, and 817.6 with 7, 2 and 5 d.f., respectively. $F_{2,5} = 2.81$, which is not significant even at the level of 10%.

Regression through the origin:

$\hat{y} = 0.9\ w = 0.495\ h$.

The S.E.s for the two slopes are 2.42 and 2.28.

The total regression and residual SS are 1737.9, 882.1, and 855.7 with 8, 2, and 6 d.f., respectively. $F_{2,6} = 3.1$, which is not significant even at the level of 10%.

The final conclusion is that neither of these two multiple regressions with the eight completed observations are preferable to the individual regressions in Example 11.4 of income on husband's age or wife's age.

11.5 As seen in Example 11.5, the proportions of the entire population and the respondents having the attribute of interest are $P = 0.41$ and $P_1 = 0.45$. Further, $W_g = (9/20, 7/20, 4/20)$ and $P_g = (51/90, 28/70, 3/40)$.

The post-stratified estimator is $\hat{P} = \Sigma W_g p_{g1}$ with variance $V(\hat{P}) = \Sigma W_g^2 V(p_{g1})$. For the first type of stratification, $P_{g1} = (36/60, 16/40, 2/20)$ and $\hat{P} = 0.43$. For the second type of stratification, $P_{g1} = (2/8, 18/20, 16/20)$ and $\hat{P} = 0.4625$. Thus, the biases of $\hat{P}$ for these two type of stratification, respectively, are $0.43 - 0.41 = 0.02$ and $0.4625 - 0.41 = 0.0525$.

Responses random:

The sample proportion p_1 obtained from the responding units is unbiased for P. Further, $V(p_1) = [(2000\ 95)/1999](.41)(.59)/95 = 0.002427$, and hence $S.E.(p_1) = 0.0493$.

For both types of poststratification, $\hat{P}$ is unbiased for P. For the first type, $V(\hat{P}) = 0.002$ and $S.E.(\hat{P}) = 0.045$. For the second type, $V(\hat{P}) = 0.00226$ and $S.E.(\hat{P}) = 0.048$.

Responses not random:

In this case, p_1 has a bias of $0.45 - 0.41 = 0.04$. Now $V(p_1) = [(1200 - 95)/1199](.45)(.55)/95 = 0.0024$ and $MSE(p_1) = 0.004$.

For the first type of stratification, $V(\hat{P}) = 0.00204$ and $MSE(\hat{P}) = 0.00244$. For the second type of stratification, $V(\hat{P}) = 0.001686$ and $MSE(\hat{P}) = 0.004443$. The first type of stratification has relatively smaller bias and MSE.

11.7 Known sizes: $N_{12} = 900$ and $N_{22} = 900$.

For the Junior–Senior class, $\bar{Y} = (.5)(12.5) + (.5)(14.4) = 13.45$. Further, $v(\hat{\bar{Y}}) = [(900 - 120)/900 \times 120](22) + [(900 - 85)/900 \times 85](32) = 0.0125$, and hence $S.E.(\hat{\bar{Y}}) = 0.354$. Estimated sizes: $\hat{N}_{12} = 3600(165/400) = 1485$ and $\hat{N}_{22} = 3600(140/400) = 1260$. Now, $\hat{\bar{Y}} = 13.374$, $v(\hat{\bar{Y}}) = 0.1235$ and $S.E.(\hat{\bar{Y}}) = 0.351$. The estimate of the mean and its S.E. do not differ much from the above case of known sizes.

Chapter 12

12.1 Deleting the ith observation, $\bar{y}_i' = (n\bar{y} - y_i)/(n - 1)$. The pseudo values are $n\bar{y} - (n - 1)\bar{y}_i' = y_i$. Since the jackknife estimator is the average of these values, it is the same as $\bar{y}$.

12.3 For this data, $\bar{x} = 6.51, \bar{y} = 10.48$, $s_x^2 = 72.56$, $s_y^2 = 114.69$, $s_{xy} = 86.896$.

From the above figures, $\hat{R} = 1.61$ and $v(\hat{R}) = 0.0542$.

Since $c_{xx} = 1.71$ and $c_{xy} = 1.27, \hat{R}_T = 1.61(1 - .044) = 1.54$.

For the jackknife, we found that $\hat{R}_J = 1.4958$ and $v_J(\hat{R}) = 0.1136$. This estimator for the ratio does not differ much from $\hat{R}$. However, $v(\hat{R})$ and $v_J(\hat{R})$ differ substantially.

12.5 We find from Table 12.1 that except for the sixth unit, the total enrollment (x), is at least five times as much as the private enrollment (y) for the remaining nine units. We have analyzed this problem with these units. Now the mean and variance of x are 174.22 and 13091. For y, the mean and variance are 32.667 and 512. The covariance is $s_{xy} = 2273$.

(a) The regression coefficient is $b = 2273/13091 = 0.1736$. Further, $\hat{\sigma}^2 = 134.26$ and $v(b) = 134.26/104728 = 0.00128$.

(b) For the jackknife, $b_J = 0.1746$ and $v(b_J) = 0.0044$.

This estimate for the regression coefficient does not differ much from the classical estimator b, but there is much difference between the variances for b obtained from these two procedures.

12.7 Denoting the numerator of $V(\hat{R})$ by V, the covariance between the pseudo values can be expressed as Cov $(\hat{R}_i', \hat{R}_j') = n^2 V/n - 2n(n-1)V/(n-1) + (n-1)^2 \ V/(n-2) = V/(n-2) = V/(n-2)$, which vanishes for large n.

References

Alho, J.M. (1990). Adjusting for nonresponse bias using logistic regression. *Biometrika*, **77**, 617–624.

Ansen, H., Hallen, S.-A., and Yalender, H. (1988). Statistics for regional and local planning in Sweden. *J. Official Stat.*, **4**, 35–46.

Bailar, B.A. and Bailar, J.C., III (1978). Comparison of two procedures for imputing missing survey values. In *Proceedings of the Survey Research Methods Section*, American Statistical Association, Washington, D.C., 462–467.

Bailar, B.A. and Bailar, J.C., III (1983). Comparison of the biases of the "Hot Deck" imputation procedure with an "Equal Weights" imputation procedure. *Incomplete Data in Sample Surveys*, W.G. Madow, I. Olkin, and D.B. Rubin, Eds., Vol. 3, Academic Press, New York, 420–470.

Bailar, B.A. and Lanphier, C.M. (1978). *Development of Survey Methods to Assess Survey Practices*. American Statistical Association, Washington, D.C.

Bailar, B.A., Bailey, L., and Corby, C. (1978). A comparison of some adjustment and weighting procedures for survey data. In *Proceedings of the Survey Research Methods Section*, American Statistical Association, Washington, D.C., 175–200.

Baker, S.G. and Laird, N.M. (1988). Regression analysis of categorical variables with outcome subject to nonignorable nonresponse. *J. Am. Stat. Assoc.*, **83**, 62–69.

Battese, G.E., Harter, R.M., and Fuller, W.A. (1988). An error component model for prediction of county crop areas using survey and satellite data. *J. Am. Stat. Assoc.*, **83**, 28–36.

Bayless, D.L. and Rao, J.N.K. (1970). An empirical study of stabilities of estimators and variance estimators in unequal probability sampling ($n = 3$ or 4). *J. Am. Stat. Assoc.*, **65**, 1645–1667.

Bean, J.A. (1975). Distribution and Properties of Variance Estimators for Complex Multistage Probability Samples. National Center for Health Statistics, Washington, D.C., Series 2, 65.

Bell, W.R. (1996). Using information from demographic analysis in post-enumeration survey estimation. *J. Am. Stat. Assoc.*, **88**(423), 1106–1118.

Bell, W.R. (1999). Accounting for uncertainty about variances in small area estimation. *Proc. Int. Stat. Inst.*, Topic 47, 57–59.

Bethel, J. (1989). Sample allocation in multivariate surveys. *Surv. Methodol.*, **15**, 47–57.

Bethlehem, J.G. (1988). Reduction of nonresponse bias through regression estimation. *J. Official Stat.*, **4**, 252–260.

Biemer, P.P., Groves, R.M., Lyberg, L.E., Mathiowetz, N.A. and Sudman, S. (1991). *Measurement Errors in Surveys*. John Wiley and Sons, New York.

Binder, D.A. (1983). On the variances of asymptotically normal estimators from complex surveys. *Int. Stat. Rev.*, **51**, 279–292.

Binder, D.A. (1991). A framework for analyzing categorical survey data with nonresponse. *J. Official Stat.*, **7**, 393–404.

Binder, D.A. and Patak, Z. (1994). Use of estimating functions for estimation from complex surveys. *J. Am. Stat. Assoc.*, **89**(427), 1035–1043.

Binder, D.A. and Theberge, A. (1988). Estimating the variance of raking-ratio estimators. *Can. J. Stat.*, **16**, 47–55.

Booth, G. and Sedransk, J. (1969). Planning some two-factor comparative surveys. *J. Am. Stat. Assoc.*, **64**, 560–573.

Booth, J.G., Buttler, R.W., and Hall, P. (1994). Bootstrap methods for finite populations. *J. Am. Stat. Assoc.*, **89**, 1282–1289.

Boteman, S.L., Massey, J.T., and Shimizu, I.M. (1982). Effect of weighting adjustments on estimates from a random-digit-dialed telephone survey. In *Proceedings of the Survey Research Methods Section*, American Statistical Association, Washington, D.C., 139–144.

Brackstone, G.J. (1998). National Longitudinal Survey of Children and Youth. *Surv. Stat.*, **39**, (13), International Statistical Institute.

Brackstone, G.J. and Rao, J.N.K. (1976). Raking ratio estimators, *Surv. Methodol.*, **2**, 63–69.

Brewer, K.R.W. (1963). Ratio estimation in finite populations. Some results deducible from the assumption of an underlying stochastic process. *Aust. J. Stat.*, **5**(3), 93–105.

Brewer, K.R.W. and Hanif, M. (1983). *Sampling with Unequal Probabilities*. Springer-Verlag, New York.

Brown, J.J., Diamond, I.D., Chambers, R.L., Buckner, L.J. and Teague, A.D. (1999). *J. R. Stat. Soc.*, **A162**(2), 247–267.

Brown, P.J., Firth, D., and Payne, C.D. (1999). Forecasting on British Election Night 1997. *J. R. Stat. Soc.*, **A162**(2), 211–226.

Buck, S.F. (1960). A method of estimation of missing values in multivariate data suitable for use with an electronic computer. *J. R. Stat. Soc.*, **B22**, 302–306.

Calvin, J.A. and Sedransk, J. (1991). Bayesian and frequentist predictive inference for the patterns of care studies. *J. Am. Stat. Assoc.*, **86**, 36–48.

Carfanga, E. (1997). Correlograms and sample designs in agriculture. *Bull. Int. Stat. Inst.*, **51**(1), 5–6.

Casady, R.J. and Valliant, R. (1993). Conditional properties of post-stratified estimators under normal theory. *Surv. Methodol.*, **19**, 183–192.

Cassel, C.M., Särndal, C.E., and Wretman, J.H. (1976). Some results on generalized difference estimation and generalized regression estimation for finite populations. *Biometrika,* **63,** 615–620.

Cassel, C.M., Särndal, C.E., and Wretman, J.H. (1977). *Foundations of Inference in Survey Sampling.* John Wiley & Sons, New York.

Cassel, C.M., Särndal, C.E., and Wretman, J.H. (1983). Some uses of statistical models in connection with the nonresponse problem. In *Incomplete Data in Sample Surveys,* Vol. 3, W.G. Madow and I. Olkin, Eds., Academic Press, New York.

Chapman, D.W., Bailey, L., and Kasprzyk, D. (1986). Nonresponse adjustment procedures at the U.S. Bureau of the Census. *Surv. Methodol.,* **12,** 161–179.

Chatterjee, S. (1968). Multivariate stratified surveys. *J. Am. Stat. Assoc.,* **63,** 530–534.

Chiu, H.Y. and Sedransk, J. (1986). A Bayesian procedure for imputing missing values in sample surveys. *J. Am. Stat. Assoc.,* **81,** 667–676.

Citro, C.F., Cohen, M.L., and Kalton, G., Eds. (1998), *Small-Area Estimates of School-Age Children in Poverty.* National Academy Press, Washington, D.C.

Cochran, W.G. (1942). Sampling theory when the sampling units are of unequal sizes. *J. Am. Stat. Assoc.,* **37,** 199–212.

Cochran, W.G. (1946). Relative accuracy of systematic and stratified random samples for a certain class of populations. *Ann. Math. Stat.,* **17,** 164–177.

Cochran (1961). Comparison of methods for stratum boundaries. *Bull. Int. Stat. Inst.,* 38(2), 345–358.

Cochran, W.G. (1977). *Sampling Techniques.* John Wiley & Sons, New York.

Cochran, W.G. (1983). Historical perspective. In *Incomplete Data in Sample Surveys,* Vol. 2, W. G. Madow, I. Olkin, and B. Rubin, Eds., Academic Press, New York, 11–25.

Cohen, S.B. and Braden, J.J. (1993). Alternative options for state level estimates in the National Medical Expenditure Survey. *Proceedings of the Survey Research Methods Section,* American Statistical Association, Washington, D.C., 364–369.

Collins, M. (1999). Sampling for U.K. telephone surveys. *J. R. Stat. Soc.,* **A162**(1), 1–4.

Conaway, M.R. (1992). The analysis of repeated categorical measurements subject to nonignorable nonresponse. *J. Am. Stat. Assoc.,* **87,** 817–824.

Cressie, N. (1989). Empirical Bayes estimation of undercount in the Decennial Census. *J. Am. Stat. Assoc.,* **84,** 1033–1044.

Cressie, N. (1992). REML estimation in empirical Bayes smoothing of census undercount. *Surv. Methodol.,* **18,** 75–94.

Dalenius, T. and Hodges, J.L., Jr. (1959). Minimum variance stratification. *J. Am. Stat. Assoc.,* **54,** 88–101.

Datta, G.S., Ghosh, M., Huang, E., Isaki, C.T., Schltz, L.K., and Tsay, J.H. (1992). Hierarchical and empirical Bayes methods for adjustment of census undercount: the 1980 Missouri dress rehearsal data. *Surv. Methodol.,* **18,** 95–108.

Deming, W.E. (1953). On a probability mechanism to attain an economic balance between the resultant error of nonresponse and the bias of nonresponse. *J. Am. Stat. Assoc.*, **48**, 743–772.

Dempster, A.P. and Raghunathan, T.E. (1987). Using covariates for small area estimation: a common sense Bayesian approach. In *Small Area Statistics*, R. Platek, J.N.K. Rao, C.E. Särndal, and M.P. Singh, Eds., John Wiley & Sons, New York.

Des Raj (1968). *Sampling Theory.* McGraw-Hill, New York.

Deville, J.-C. and Särndal, C.E. (1992). Calibration estimator in survey sampling. *J. Am. Stat. Assoc.*, **87**, 376–382.

Dillman, D. (1978). *Mail and Telephone Surveys.* John Wiley & Sons, New York.

Ding, Y. and Fienberg, S.E. (1994). Dual system estimation of census undercount in the presence of matching error. *Surv. Methodol.*, **20**, 149–158.

Dorfman, A.H. (1994). A note on variance estimation for the regression estimator in double sampling. *J. Am. Stat. Assoc.*, **89**, 137–140.

Drew, J.D., Singh, M.P., and Choudhry, G.H. (1982). Evaluation of small area techniques for the Canadian Labor Force Survey. *Surv. Methodol.*, **8**, 17–47.

Dumicic/, K. and Dumicic/, S. (1999). The sample strategy for an UNICEF survey in Croatia in 1996. *Proc. Int. Stat. Inst.*, **D**, 287–288.

Duncan, G.J. and Kalton, G. (1987). Issues of design and analysis of surveys across time. *Int. Stat. Rev.*, **55**, 97–117.

Durbin, J. (1959). A note on the application of Quenouille's method of bias reduction to the estimation of ratios. *Biometrika*, **46**, 477–480.

Durbin, J. (1967). Design of multi-stage surveys for the estimation of sampling errors. *Appl. Stat.*, **16**, 152–164.

Dutka, S. and Frankel, L.R. (1995). Probability sampling for marketing research surveys in 56 countries. *Proc. Int. Assoc. Survey Statisticians*, 97–117.

Edler, L., Pilz, L.R., and Potschke-Langer, M. (1999). On the prevalence of smoking among young school children and its association with risky behaviour and health risks. *Proc. Int. Stat. Inst.*, **E**, 293–294.

Efron, B. (1982). *The Jackknife, the Bootstrap and Other Resampling Plans*, Society for Industrial and Applied Mathematics, Philadelphia.

Efron, B. (1994). Missing data, imputation and the bootstrap. *J. Am. Stat. Assoc.*, **89**, 463–479.

Efron, B. and Tibshirani, R.J. (1993). *An Introduction to the Bootstrap.* Chapman & Hall, New York.

Ekman, G. (1959). An approximation used in univariate stratification. *Ann. Math. Stat.*, **30**, 219–229.

Ericksen, E.P. (1974). A regression method for estimating populations of local areas. *J. Am. Stat. Assoc.*, **69**, 867–875.

Ericksen, E.P. and Kadane, J.B. (1985). Estimating the population in a census year (with comments and rejoinder). *J. Am. Stat. Assoc.*, **80**, 98–131.

Ericksen, E.P., Kadane, J.B., and Tukey, J.W. (1989). Adjusting the 1980 census of population and housing. *J. Am. Stat. Assoc.*, **84**, 927–944.

Ericson, W.A. (1967). Optimal sample design with nonresponse. *J. Am. Stat. Assoc.*, **69**, 63–78.

Ericson, W.A. (1969). Subjective Bayesian models in sampling finite populations. *J. R. Stat. Soc.,* **B31,** 195–233.

Ernst, L.R. (1978). Weighting to Adjust for Partial Nonresponse. In *Proceedings of the Social Statistics Section,* American Statistical Association, Washington, D.C., 468–473.

Ernst, L.R. (1980). Variance of the estimated mean for several imputation procedures. In *Proceedings of the Social Statistics Section,* American Statistical Association, Washington, D.C., 716–720.

Falrosi, P.D., Falrosi, S., and Russo, A. (1994). Empirical comparison of small area estimation methods for the Italian Labor Force Survey. *Surv. Methodol.,* **20,** 171–176.

Fay, R.E. (1986). Causal models for patterns of nonresponse. *J. Am. Stat. Assoc.,* **81,** 354–365.

Fay, R.E. (1996). Alternative paradigms for the analysis of imputed survey data. *J. Am. Stat. Assoc.,* **91,** 480–490.

Fay, R.E. and Harriet, R.A. (1979). Estimates of income for small places: an application of James-Stein procedures to census data. *J. Am. Stat. Assoc.,* **74,** 269–277.

Fay, R.E. and Train, G.F. (1995). Aspects of survey and model-based postcensal estimation of income and poverty characteristics for states and counties. *Proceedings of the Section of Government Statistics,* American Statistical Association, Washington, D.C.

Fellegi, I. (1963). Sampling with varying probabilities without replacement: rotating and non-rotating samples. *J. Am. Stat. Assoc.,* **58,** 183–201.

Fellegi, I.P. and Holt, D. (1976). A systematic approach to automatic edit and imputation. *J. Am. Stat. Assoc.,* **71,** 17–35.

Ferranti, M.R. and Pacei, S. (1997). Small area estimation of agricultural production. The Italian Survey on Farms. *Bull. Int. Stat. Inst.,* **LVII**(1), 511–512.

Fieller, E.C. (1932). The distribution of the index in a normal bivariate population, *Biometrika,* **24,** 428–440.

Fienberg, S.E. (1992). Bibliography on capture-recapture modeling with application to census undercount adjustment. *Surv. Methodol.,* **18,** 143–154.

Fisher, R. and Siegel, (1997). Methods used for small area poverty and income estimation. *Proceedings of the Social Statistics Section,* American Statistical Association, Washington, D.C.

Fuller, W.A. and Isaki, C.T. (1981). Survey design under superpopulation models. In *Current Topics in Survey Sampling,* D. Krewski, J.N.K. Rao, and R. Platek, Eds., Academic Press, New York, 199–226.

Fuller, W.A., Loughlin, M.E., and Baker, H.D. (1994). Regression weighting in the presence of nonresponse with application to the 1987–1988 Nationwide Food Consumption Survey. *Surv. Methodol.,* **20,** 75–85.

Gelman, A., Carlin, J.B., Stern, H.S., and Rubin, D.B. (1995). *Bayesian Data Analysis.* Chapman & Hall/CRC Press, New York.

Gelman, A., King, G., and Liu, C. (1998). Not asked and not answered: multiple imputation for multiple surveys. *J. Am. Stat. Assoc.,* **93,** 846–857.

Ghosh, M. and Lahiri, P. (1987). Robust Empirical Bayes estimation of variances from stratified samples. *Sankhya,* **B49,** 78–89.

Ghosh, M. and Meeden, G. (1986). Empirical Bayes' estimation in finite populations. *J. Am. Stat. Assoc.,* **81,** 1058–1062.

Ghosh, M. and Rao, J.N.K. (1994). Small area estimation: an appraisal. *Stat. Sci.,* **8,** 76–80.

Giles, P. (1988). A model for generalized edit and imputation of survey data. *Can. J. Stat.,* **16S,** 57–73.

Glynn, R.J., Laird, N.M., and Rubin, D.B. (1993). Multiple imputation in mixture models for nonignorable nonresponse with follow-ups. *J. Am. Stat. Assoc.,* **88,** 984–993.

Godambe, V.P. (1955). A unified theory of sampling from finite population. *J. R. Stat. Soc.,* **B17,** 269–278

Godambe, V.P. (1966). A new approach to sampling from finite populations I,II. *J. R. Stat. Soc.* **B28**(2), 310–328.

Godambe, V.P. and Thompson, M.E. (1986). Some optimality results in the presence of nonresponse. *Surv. Methodol.,* **12,** 29–36 (correction, **13,** 123).

Gonzalez, M.E. (1973). Use and evaluation of synthetic estimates. In *Proceedings of the Social Sciences Section,* American Statistical Association, Washington, D.C., 33–36.

Gover, A.R. (1979). Nonresponse in the Canadian Labour Force Survey. *Surv. Methodol.,* **5,** 29–58.

Griffith, J.E. and Frase, M.J. (1995). Longitudinal studies in the United States: tracking students to understand educational progress. *Bull. Int. Stat. Inst.,* 50th Session, **LVI**(3), 1035–1052.

Gross, S. (1980). Median estimation in sample surveys. In *Proceedings of the Section on Survey Research Methods,* American Statistical Association, Washington, D.C., 181–184.

Groves, R.M. and Nicholls, W.L., II (1986). The status of computer-assisted telephone interviewing (1986): Part II—Data quality issues. *J. Official Stat.,* **2,** 117–134.

Groves, R.M., Biemer, P.P., Lyberg, L.E., Massey, J.T., Nicholls, W.L., and Waksberg, J., Eds. (1988), *Telephone Survey Methodology,* John Wiley & Sons, New York.

Gurney, M. and Jewett, R.S. (1975). Constructing orthogonal replications for variance estimation. *J. Am. Stat. Assoc.,* **70,** 819–821.

Hanif, M. and Brewer, K.R.W. (1980). Sampling with unequal probabilities without replacement: a review. *Int. Stat. Rev.,* **48,** 317–335.

Hansen, M.H. and Hurwitz, W.N. (1946). The problem of nonresponse in sample surveys. *J. Am. Stat. Assoc.,* **41,** 517–529.

Hansen, M.H., Hurwitz, W.N. and Gurney, M. (1946). Problems and methods of the sample survey of business. *J. Am. Stat. Assoc.,* **41,** 173–189.

Hartley, H.O. and Ross, A. (1954). Unbiased ratio estimates. *Nature,* **174,** 270–271.

Heitjan, D.F. and Little, R.J.A. (1988). Multiple imputation for the fatal accident system. In *Proceedings of the Section on Survey Research Methods*, American Statistical Association, Washington, D.C., 93–102.

Hendricks, W.A. (1949). Adjustment for bias by nonresponse in mailed surveys. *Agric. Econ. Res.*, **1**, 52–56.

Herzog, T.N. and Rubin, D.B. (1983). Using multiple imputations to handle nonresponse in sample surveys. In *Incomplete Data in Sample Surveys*, Vol. 2, W. G. Madow, I. Olkin, and D.B. Rubin, Eds., Academic Press, New York, 210–245.

Hidiroglou, M.A. and Särndal, C.E. (1985). An empirical study of some regression estimators for small domains. *Surv. Methodol.*, **11**, 65–77.

Hogan, H. (1993). The 1990 post-enumeration survey: operations and results. *J. Am. Stat. Assoc.*, **88**, 1047–1060.

Holt, D. and Holmes, D.J. (1994). Small domain estimation for unequal probability survey designs. *Surv. Methodol.*, **20**, 23–31.

Holt, D., Smith, T.M.F., and Tomberlin, T.J. (1979). A model-based approach to estimation for small subgroups of a population. *J. Am. Stat. Assoc.*, **74**, 405–410.

Holt, D., Smith, T.M.F., and Winter, P.O. (1980). Regression analysis of data from complex surveys. *J. R. Stat. Soc.*, **A143**, 474–487.

Horvitz, D.G. and Thompson, D.J. (1952). A generalization of sampling without replacement from a finite universe, *J. Am. Stat. Assoc.*, **47**, 663–685.

Iachan, R. and Dennis, M.L. (1993). A multiple frame approach to sampling the homeless and transient population. *J. Official Stat.*, **9**, 747–764.

Isaki, C.T. and Fuller, W.A. (1982). Survey design under the regression superpopulation model, *J. Am. Stat. Assoc.*, **77**, 89–96.

Isaki, C.T., Schultz, L.K., Smith, P.J. and Diffendal, G.J. (1987). Small area estimation research for census undercount-Progress report. In, *Small Area Statistics*, R. Platek, J.N.K. Rao, C.E. Sarndal and M.P. Singh, Eds., John Wiley and Sons, New York, 219–238.

Jackson, J.E. and Rao, P.S.R.S. (1983). Estimation procedures in the presence of nonresponse. In *Proceedings of the Section on Survey Research Methods*, American Statistical Association, Washington, D.C., 273–276.

Jacobs, E., Jacobs, C., and Dippo, C. (1989). The U.S. Consumer Expenditure Survey. *Bull. Int. Stat. Inst.*, **LIII**(2), 123–142.

Jagers, P. (1986). Post-stratification against bias in sampling. *Int. Stat. Rev.*, **54**, 159–167.

Jones, H.L. (1974). Jackknife estimation of functions of stratum means. *Biometrika*, **61**, 343–348.

Kadane, J.B. (1993). Subjective Bayesian analysis for surveys with missing data. *Statistician*, **42**, 415–426.

Kalton, G. and Kasprzyk, D. (1986). The treatment of missing survey data. *Surv. Methodol.*, **12**, 1–16.

Kalton, G. and Kish, L. (1984). Some efficient random imputation methods. *Commun. Stat.*, **A13**(16), 1919–1940.

Katzoff, M.J., Jones, G.K., Curtin, L.R., and Graubard, B. (1989). Two empirical studies of statistical methods applied to data from complex surveys. In *Proceedings of the Survey Research Methods Section, American Statistical Association*, Washington, D.C., 757–762.

Kaufman, G.H. and King, B. (1973). A Bayesian analysis of nonresponse in dichotomous process. *J. Am. Stat. Assoc.*, **68**, 670–678.

Kish, L. (1965). *Survey Sampling*. John Wiley & Sons, New York.

Kish, L. and Frankel, M.R. (1974). Inference from complex samples. *J. R. Stat. Soc.*, **B36**, 1–37.

Kott, P.S. (1990a). Survey Processing System (SRS) Summary. *National Agricultural Statistics Service (NASS) Staff Report*, No. SRB-90-08.

Kott, P.S. (1994b). A note on handling nonresponse in sample surveys. *J. Am. Stat. Assoc.*, **89**, 693–696.

Kott, P.S. (1994c). Reweighting and variance estimation for the characteristics of business owners survey. *J. Official Stat.*, **10**, 407–418.

Kovar, J.G., Rao, J.N.K., and Wu, C.F.J. (1988). Bootstrap and other methods to measure errors in survey estimates. *Can. J. Stat.*, **16**, 25–45.

Krewski, D. and Chakrabarti, R.P. (1981). On the stability of the jackknife variance estimator in ratio estimation. *J. Stat. Planning Inference*, **5**, 71–79.

Krewski, D. and Rao, J.N.K. (1981). Inference from stratified samples: properties of the linearization, jackknife and balanced repeated replication methods. *Ann. Stat.*, **9**, 1010–1019.

Lahiri, D.B. (1951). A method of sample selection providing unbiased ratio estimates. *Bull. Int. Stat. Inst.*, **33**(2), 133–140.

Lahiri, P. and Rao, J.N.K. (1995). Robust estimation of mean square error of small area estimation. *J. Am. Stat. Assoc.*, **90**(430), 758–766.

Lehtonen, R. and Veijanen, A. (1999). Multilevel-model assisted generalized regression estimators for domain estimation. *Proc. Int. Stat. Inst.*, **L**, 227–228.

Lehtonen, R. and Pahkinen, E.J. (1995). *Practical Methods for Design and Analysis of Complex Surveys*. John Wiley and Sons, New York.

Lemeshow, S. and Levy, P. (1978). Estimating the variances of the ratio estimates in complex surveys with two primary units per stratum—a comparison of balanced replication and jackknife techniques. *J. Stat. Comp. and Simul.*, **8**, 191–205.

Lessler, J.T. and Kalsbeek, W.D. (1992). *Nonsampling Errors in Surveys*. John Wiley & Sons, New York.

Liao, H. and Sedransk, J. (1983). Selection of strata sample sizes for the comparison of domain means. *J. Am. Stat. Assoc.*, **78**, 870–878.

Lipsitz, S.R., Laird, N.M., and Harrington, D.P. (1994). Weighted least squares analysis of repeated categorical measurements with outcomes subject to nonresponse. *Biometrics*, **50**, 1102–1116.

Little, R.J.A. (1982). Models for nonresponse in sample surveys. *J. Am. Stat. Assoc.*, **77**, 237–250.

Little, R.J.A. (1986). Survey nonresponse adjustment for estimates of means. *Int. Stat. Rev.*, **54**(2), 139–157.

Little, R.J.A. (1995). Modelling the drop-out mechanism in repeated measures studies. *J. Am. Stat. Assoc.,* **90,** 1112–1121.

Little, R.J.A. and Rubin, D.B. (1987). *Statistical Analysis with Missing Data.* John Wiley & Sons, New York.

Lohr, S.L. and Rao, J.N.K. (2000). Inference from dual frame surveys. *J. Am. Stat. Assoc.,* **95**(449), 271–280, 1645–1667.

Longford, N.T. (1999). Multivariate shrinkage estimation of small area means and proportions. *J. R. Stat. Soc.,* **A162**(2), 227–246.

Lundstrom, S. (1987). An evaluation of small area estimation methods. The case of estimating the number of unmarried cohabiting persons in Swedish municipalities. In *Small Area Statistics,* R. Platek, J.N.K. Rao, C.E. Särndal, and M. P. Singh, Eds., John Wiley & Sons, New York.

Madow, W.G. and Madow, L.H. (1944). On the theory of systematic sampling. *Ann. Math. Stat.,* **15,** 1–24.

Mahalanobis, P.C. (1944). On large-scale sample surveys. *Philos. Trans. R. Stat. Soc.,* **B231,** 329–451.

Mahalanobis, P.C. (1946). Recent experiments in statistical sampling in the Indian Statistical Institute. *J. R. Stat. Soc.,* **109,** 325–370.

Malec, D. and Sedransk, J. (1985). Bayesian inference for finite population parameters in multistage cluster sampling. *J. Am. Stat. Assoc.,* **80,** 897–902.

Malec, D., Sedransk, J., Moriarity, C.L., and LeClere, F.B. (1997). Small area inference for binary variables in the National Health Interview Survey. *J. Am. Stat. Assoc.,* **92,** 815–826.

Marker, D.A. (1993). Small area estimation for the U.S. National Health Interview Survey. *Proceedings of the Survey Research Methods Section,* American Statistical Association, Washington, D.C., 11–20.

Martin, J., O'Muircheartaigh, C., and Curtice, J. (1993). The use of CAPI for attitude surveys: an experimental comparison with traditional methods. *J. Official Stat.,* **9,** 641–661.

Massey, J.T., Moore, T.F., Parsons, V.L., and Tadros, W. (1989). Design and Estimation for the National Health Interview Survey, 1985–94. Vital and Health Statistics, Series 2, No. 110. DHHS Publication No. PHS 89–1384. Public Health Service, U.S. Government Printing Office, Washington, D.C.

McCarthy, P.J. (1966). Replication: An Approach to the Analysis of Data from Complex Surveys. National Center for Health Statistics, Washington, D.C., Series 2, 14.

McCarthy, P.J. (1969). Pseudoreplication: half-samples. *Rev. Int. Stat. Inst.,* **37,** 239–264.

McCarthy, P.J. and Snowden, C.B. (1985). The Bootstrap and Finite Population Sampling. Vital and Health Statistics, Series 2 (95), Public Health Service Publication 85–1369, U.S. Government Printing Office, Washington, D.C.

Midzuno, H. (1951). On the sampling system with probability proportionate to sum of sizes. *Ann. Inst. Stat. Math.,* **2,** 99–108.

Mosteller, C.F., Hyman, H., McCarthy, P., Marks, E.S., and Truman, D.B. (1949). The Pre-election Polls of 1948, Bull. 60, Social and Scientific Research Council, New York.

Murthy, M.N. (1957). Ordered and unordered estimators in sampling without replacement. *Sankhya*, **18**, 379–390.

Murthy, M.N. (1967). *Sampling Theory and Methods.* Statistical Publishing Society, Calcutta, India.

Narain, R.D. (1951). On sampling without replacement with varying probabilities. *J. Ind. Soc. Agri. Stat.*, **3**, 169–174.

Narain, P. and Srivastava, A.K. (1995). Census and survey issues in Indian Agricultural Census. In *Proceedings of the International Association of Survey Statisticians*, International Statistical Institute, 229–239.

Nathan, G. (1988). Inference based on data from complex sample designs. In *Handbook of Statistics,* Vol. **6**, P.R. Krishnaiah and C.R. Rao, Eds., North Holland, New York, 247–266.

Neyman, J. (1934). On the two different aspects of the representative sampling and the method of purposive selection. *J. R. Stat. Soc.*, **97**, 558–606.

Neyman, J. (1938). Contribution to the theory of sampling human populations. *J. Am. Stat. Assoc.*, **33**, 101–116.

Nordberg, L. (1989). Generalized linear modeling of sample survey data. *J. Official Stat.*, **5**, 223–239.

Oh, H.L. and Scheuren, F.J. (1983). Weighting adjustment for nonresponse, In *Incomplete Data in Sample Surveys,* Vol. 2, W.G. Madow, I. Olkin, and B. Rubin, Eds., Academic Press, New York, 185–206.

Oh, H.L. and Scheuren, F.J. (1987). Modified raking ratio estimator, *Surv. Methodol.*, **13**, 209–219.

Olkin, I. (1958). Multivariate ratio method of estimation for finite populations, *Biometrika*, **45**, 154–165.

O'Muircheartaigh, C.A. (1977). Proximum designs for crude sampling frames. In *Proceedings of the 41st Session,* International Statistical Institute, **57**, 82–100.

Pathak, P.K. (1962) On sampling with unequal probability sampling. *Sankhya,* **A24**, 315–326.

Pfeffermann, D. and Burck, l. (1990). Robust small area estimation combining time series and cross-sectional data. *Surv. Methodol.*, **16**, 217–237.

Pfeffermann, D. and Nathan, G. (1985). Problems in model selection identification based on data from complex surveys. *Bull. Int. Stat. Inst.*, **LI**(2), 12.2, 1–18.

Platek, R. and Gray, B.G. (1983). Imputation methodology: total survey error. In *Incomplete Data in Sample Surveys,* W.G. Madow, I. Olkin, and D.B. Rubin, Eds., Academic Press, New York, 247–333.

Politz, A. and Simmons, W. (1949). An attempt to get the "not at home"s into the sample without callbacks. *J. Am. Stat. Assoc.*, **44**, 9–31.

Potthoff, R.F., Manton, K., and Woodbury, M.A. (1993). Correcting for nonavailability bias in surveys with follow-ups. *J. Am. Stat. Assoc.*, **88**, 1197–1207.

Prasad, N.G.N. and Rao, J.N.K. (1990). The estimation of the mean square error of small-area estimators. *J. Am. Stat. Assoc.,* **85**, 163–171.

Purcell, N.J. and Kish, L. (1980). Postcensal estimates for local areas (or domains). Int. Stat. Rev., **48**, 3–18.

Quenouille, M.H. (1956). Notes on bias in estimation. *Biometrika,* **43**, 353–360.

Raghunathan, T.E. and Grizzle, J.E. (1995). A split questionnaire survey design. *J. Am. Stat. Assoc.,* **90**, 55–63.

Rancourt, E., Lee, H. and Sarndal, C.E. (1994). Bias corrections for survey estimates from data with ratio imputed values for confounded nonresponse. *Surv. Methodol.,* **20**, 137–147.

Rao, C.R., Pathak, P.K., and Koltchinski, V.I. (1997). Bootstrap by sequential sampling. *J. Stat. Planning and Inference,* **64**, 257–281.

Rao, J.N.K. (1965). On two simple schemes of unequal probability sampling without replacement. *J. Indian Stat. Assoc.,* **3**, 173–180.

Rao, J.N.K. (1973). On double sampling for stratification and analytical surveys. *Biometrika,* **60**, 125–133 (correction, **60**, 669).

Rao, J.N.K. (1996). On variance estimation with imputed survey data. *J. Am. Stat. Assoc.,* **91**, 499–505.

Rao, J.N.K. and Bayless, D.L. (1969). An empirical study of the stabilities of estimators and variance estimators in unequal probability sampling of two units per stratum. *J. Am. Stat. Assoc.,* **64**, 540–559.

Rao, J.N.K. and Ghangurde, P.D. (1972). Bayesian optimization in sampling finite populations. *J. Am. Stat. Assoc.,* **67**, 539–553.

Rao, J.N.K. and Scott, A.J. (1981). The analysis of categorical data from complex surveys: chi-squared tests for goodness of fit and independence in two way table. *J. Am. Stat. Assoc.,* **76**, 221–230.

Rao, J.N.K. and Shao, J. (1992). Jackknife variance estimation with survey data under hot deck imputation. *Biometrika,* **79**, 811–822.

Rao, J.N.K. and Webster, J.T. (1966). On two methods of bias reduction in the estimation of ratios. *Biometrika,* **53**, 571–577.

Rao, J.N.K. and Wu, C.F.J. (1985). Inference from stratified samples: Second-order analysis of three methods for nonlinear statistics *J. Am. Stat. Assoc.,* **80**, 620–630.

Rao, J.N.K. and Wu, C.F.J. (1988). Resampling inference with complex survey data, *J. Am. Stat. Assoc.,* **83**, 231–241.

Rao, J.N.K. and Yu, M. (1992). Small area estimation by combining time series and cross-sectional data. *Proceedings of the Survey Research Methods Section,* American Statistical Association, Washington, D.C., 1–9.

Rao, J.N.K., Hartley, H.O., and Cochran, W.G. (1962). A simple procedure of unequal probability sampling without replacement. *J. R. Stat. Soc.,* **B24**, 482–491.

Rao, J.N.K., Wu, C.F.J., and Yue, K. (1992). Some recent work on resampling methods for complex surveys. *Surv. Methodol.,* **18**, 209–217.

Rao, P.S.R.S. (1968). On three procedures of sampling from finite populations. *Biometrika,* **55**, 438–440.

Rao, P.S.R.S. (1969). Comparison of four ratio-type estimates. *J. Am. Stat. Assoc.*, **64**, 574–580.

Rao, P.S.R.S. (1972). On two phase regression estimators. *Sankhya*, **A33**, 473–476.

Rao, P.S.R.S. (1974). Jackknifing the ratio estimator. *Sankhya*, **C36**, 84–97.

Rao, P.S.R.S. (1975a). Hartley-Ross type estimator with two phase sampling, *Sankhya*, **C37**, 140–146.

Rao, P.S.R.S. (1975b). On the two-phase ratio estimator in finite populations. *J. Am. Stat. Assoc.*, **70**, 839–845.

Rao, P.S.R.S. (1979). On applying the jackknife procedure to the ratio estimator. *Sankhya*, **C41**, 115–126.

Rao, P.S.R.S. (1981a). Efficiencies of nine two-phase ratio estimators for the mean. *J. Am. Stat. Assoc.*, **76**, 434–442.

Rao, P.S.R.S. (1981b). Estimation of the mean square error of the ratio estimator. In *Current Topics in Survey Sampling*, D. Krewski, R. Platek, and J.N.K. Rao, Eds., Academic Press, New York, 305–315.

Rao, P.S.R.S. (1983a). Randomization approach. In *Incomplete Data in Sample Surveys*, Vol. 2, W.G. Madow, I. Olkin, and D.B. Rubin, Eds., Academic Press, New York, 97–105.

Rao, P.S.R.S. (1983b). Hansen-Hurwitz method for subsampling the nonrespondents. In *Encyclopedia of Statistical Sciences*, Vol. 3, S. Kotz and N.L. Johnson, Eds., John Wiley & Sons, New York, 573–574.

Rao, P.S.R.S. (1983c). Callbacks, follow-ups and repeated telephone calls. In *Incomplete Data in Sample Surveys*, Vol. 2, W.G. Madow, I. Olkin, and D.B. Rubin, Eds., Academic Press, New York, 33–44.

Rao, P.S.R.S. (1986). Ratio estimation with subsampling the nonrespondents. *Surv. Methodol.*, **12**(2), 217–230.

Rao, P.S.R.S. (1987). Ratio and regression estimators. *Handbook of Statistics*, Vol. **6**, P. Krishnaiah and C.R. Rao, Eds., Academic Press, New York, 449–468.

Rao, P.S.R.S. (1990). Regression estimators with subsampling the nonrespondents. In *Theory and Pragmatics of Data Quality Control*, G.E. Liepins and V.R.R. Uppuluri, Eds., Marcel Dekker, New York, 191–208.

Rao, P.S.R.S. (1991). Ratio estimators with unequal probability sampling. *Bull. Int. Stat. Inst.*, **48**(2), 542–543.

Rao, P.S.R.S. (1997). *Variance Components Estimation: Mixed Models, Methodologies and Applications*. Chapman & Hall/CRC Press, Boca Raton, FL.

Rao, P.S.R.S. (1998a). Ratio estimators (update). In *Encyclopedia of Statistical Sciences*, Update Vol. 2, S. Kotz, C.B. Read, and D.L. Banks, Eds., John Wiley & Sons, New York, 570–575.

Rao, P.S.R.S. (1998b). Double Sampling. In *Encyclopedia of Biostatistics*, P. Armitage and T. Colton, Eds., John Wiley & Sons, New York.

Rao, P.S.R.S. and Katzoff, M.J. (1996). Bootstrap for finite populations. *Commun. Stat. Simulation Computation*, **25**(4), 979–994.

Rao, P.S.R.S. and Rao, J.N.K. (1971). Small sample results for ratio estimators, *Biometrika*, **58**, 625–630.

Royall, R.M. (1970). On the finite population sampling theory under certain linear regression models, *Biometrika,* **57,** 377–387.

Royall, R.M. (1976). The linear least-squares prediction approach to two-stage sampling. *J. Am. Stat. Assoc.,* **71,** 657–664 (correction, **74,** 516).

Royall, R.M. (1986). The prediction approach to robust variance estimation to two-stage cluster sampling. *J. Am. Stat. Assoc.,* **81,** 119–123.

Rubin, D.B. (1978). Multiple imputations in sample surveys—a phenomenological Bayesian approach to nonresponse. In *Proceedings of the Survey Research Methods Section,* American Statistical Association, Washington, D.C., 20–34.

Rubin, D.B. (1979). Illustrating the use of multiple imputation to handle nonresponse in sample surveys. *Bull. Int. Stat. Inst.,* **48**(2), 517–532.

Rubin, D.B. (1986). Basic ideas of multiple imputation for nonresponse. *Surv. Methodol.,* **12,** 37–47.

Rubin, D.B. (1987). *Multiple Imputation for Nonresponse in Surveys.* John Wiley & Sons, New York.

Rubin, D.B. (1996). Multiple imputation after 18+ years. *J. Am. Stat. Assoc.,* **91,** 507–510.

Rubin, D.B. and Schenker, N. (1986). Multiple imputation for interval estimation from simple random samples with ignorable nonresponse. *J. Am. Stat. Assoc.,* **81,** 366–374.

Rubin, D.B., Schafer, J.L., and Schenker, N. (1988). Imputation strategies for missing values in postenumeration surveys. *Surv. Methodol.,* **14,** 209–221.

Samford, M.R. (1967). On sampling without replacement with unequal probabilities of selection. *Biometrika,* **54,** 499–513.

Santos, E. (1995). Retail trading surveys in Latin America. *Bull. Int. Stat. Inst.,* **LVI**(3), 889–903.

Sarndal, C.E. (1984). Design consistent versus model dependent estimation for small domains. *J. Am. Stat. Assoc.,* **79,** 624–631.

Särndal, C.E. (1986). A regression approach to estimation in the presence of nonresponse. *Surv. Methodol.,* **12,** 207–215.

Särndal, C.E. and Hidiroglou, M.A. (1989). Small domain estimation: a conditional analysis, *J. Am. Stat. Assoc.,* **84,** 266–275.

Särndal, C.E. and Swensson, B. (1985). Incorporating nonresponse modelling in a general randomization theory approach. *Proc. Int. Stat. Inst.,* 45th session.

Särndal, C.E., Swensson, B., and Wretman, J. (1992). *Model Assisted Survey Sampling.* Springer-Verlag, New York.

Schafer, J.L. (1997). *Analysis of Incomplete Multivariate Data,* Chapman & Hall, London.

Schafer, J.L. and Schenker, N. (2000). Inference with imputed conditional mean. *J. Am. Stat. Assoc.,* **95,** 144–154.

Schaible, W.L., Brock, D.B., and Schank, G.A. (1977). An empirical comparison of the simple inflation, synthetic and composite estimators for small area statistics. In *Proceedings of the Social Statistics Section,* American Statistical Association, Washington, D.C., 1017–1021.

Schucany, W.R., Gray, H.L., and Owen, D.B. (1971). On bias reduction in estimation. *J. Am. Stat. Assoc.*, **66**, 524–533.

Sedransk, J. (1965). A double sampling scheme for analytical surveys. *J. Am. Stat. Assoc.*, **60**, 985–1004.

Sedransk, J. (1967). Designing some multi-factor analytical studies. *J. Am. Stat. Assoc.*, **62**, 1121–1139.

Sen, A.R. (1953). On the estimate of variance in sampling with varying probabilities. *J. Indian Soc. Agric. Stat.*, **5**, 119–127.

Shah, B.V., Holt, M.M., and Folsom, R.E. (1977). Inference about regression models from sample survey data. *Bull. Int. Stat. Inst.*, **47**, 43–57.

Shao, J. and Sitter, R.R. (1996). Bootstrap for imputed survey data. *J. Am. Stat. Assoc.*, **91**, 1278–1288.

Shao, J. and Steel, P. (1999). Variance estimation for survey data with composite imputation and nonnegligible sampling fraction. *J. Am. Stat. Assoc.*, **94**(445), 254–265.

Shao, J. and Tu, D. (1995). *The Jackknife and the Bootstrap*. Springer-Verlag, New York.

Simmons, W.R. (1954). A plan to account for "not-at-homes" by combining weighting and callbacks. *J. Marketing*, **11**, 42–53.

Sirken, M.G. (1983). Handling missing data by network sampling. In *Incomplete Data in Sample Surveys*, Vol. 2, W.G. Madow, I. Olkin, and D.B. Rubin, Eds., Academic Press, New York, 81–89.

Sirken, M.G. and Casady, R.J. (1982). Nonresponse in dual frame surveys based on area; list and telephone frames. in *Proceedings of the Survey Research Methods Section*, American Statistical Association, Washington, D.C., 151–153.

Sitter, R.R. (1992). Resampling procedure for complex survey data. *J. Am. Stat. Assoc.*, **87**, 755–765.

Sitter, R.R. (1997). Variance estimation for the regression estimator in two-phase sampling. *J. Am. Stat. Assoc.*, **92**(438), 780–787.

Skinner, C.J. and Rao, J.N.K. (1996). Estimation in dual frame surveys with complex designs. *J. Am. Stat. Assoc.*, **91**, 349–356.

Skinner, C.J., Holt, D., and Smith, T.M.F., Eds. (1989). *Analysis of Complex Survey Data*. John Wiley & Sons, New York.

Smouse, E.P. (1982). Bayesian estimation of a finite population total using auxiliary information in the presence of nonresponse. *J. Am. Stat. Assoc.*, **77**, 97–102.

Srinath, K.P. (1971). Multiphase sampling in nonresponse problems. *J. Am. Stat. Assoc.*, **66**, 583–586.

Stasny, E.A. (1986). Estimating gross flows using panel data with nonresponse. An example from the Canadian Labor Force Survey. *J. Am. Stat. Assoc.*, **81**, 42–47.

Stasny, E.A. (1987). Some Morkov-chain models for nonresponse in estimating gross labor force flows. *J. Official Stat.*, **3**, 359–373.

Stasny, E.A. (1990). Symmetry in flows among reported victimization classifications with nonresponse. *Surv. Methodol.*, **16**, 305–330.

Stasny, E.A. (1991). Heirarchical models for the probabilities of a survey classification and nonresponse. An example from the National Crime Survey. *J. Am. Stat. Assoc.,* **86,** 296–303.

Stasny, E.A., Goel, P.K., and Rumsey, D.J. (1991). County estimates of wheat production. *Surv. Methodol.,* **17,** 211–225.

Stroud, T.W.F. (1987). Bayes and Empirical Bayes approaches to small area estimation. In *Small Area Statistics,* R. Platek, J.N.K. Rao, C.E. Sarndal and M.P. Singh, Eds., John Wiley and Sons, New York, 124–137.

Tam, S.M. (1986). Characterization of best model-based predictors in survey sampling. *Biometrika,* **73,** 232–235.

Tam, S.M. (1995). Optimal and robust strategies for cluster sampling. *J. Am. Stat. Assoc.,* **90,** 379–382.

Tin, M. (1965). Comparison of some ratio estimators. *J. Am. Stat. Assoc.,* **60,** 294–307.

Titterington, D.M. and Sedransk, J. (1986). Matching and linear regression adjustment in imputation and observational studies. *Sankhya,* **B48,** 347–367.

Treder, R.P. and Sedransk, J. (1993). Double sampling for stratification. *Surv. Methodol.,* **19,** 95–101.

Tukey, J.W. (1958). Bias and confidence in not-quite large samples. (abstr.). *Ann. Math. Stat.,* **29,** 614.

Valliant, R. (1987). Generalized variance functions in stratified two-stage sampling. *J. Am. Stat. Assoc.,* **82,** 499–508.

Valliant, R. (1993). Poststratification and conditional variance estimation. *J. Am. Stat. Assoc.,* **88,** 89–96.

Verma, V. and Thanh, L.E. (1995). Sampling errors for the demographic and health surveys program. *Proc. Int. Stat. Inst.,* Topic 3, 55–75.

Verma, V., Scott, C., and O'Muircheartaigh, C. (1980). Sample designs and sampling errors for the World Fertility Survey. *J. R. Stat. Soc.,* **A143,** 431–473

Waksberg, J. (1978). Sampling methods for random-digit dialing. *J. Am. Stat. Assoc.,* **78,** 40–46.

Wang, R., Sedransk, J., and Jinn, J.H. (1992). Secondary data analysis when there are missing observations. *J. Am. Stat. Assoc.,* **87,** 952–961.

Warner, S.L. (1965). Randomized response: a survey technique for eliminating evasive answer bias. *J. Am. Stat. Assoc.,* **60,** 63–69.

Wolter, K.M. (1979). Composite estimation in finite populations. *J. Am. Stat. Assoc.,* **74,** 604–613.

Wolter, K.M. (1985). *Introduction to Variance Estimation.* Springer-Verlag, New York.

Wolter, K.M. and Causey, B.D. (1991). Evaluation of procedures for improving population estimates for small areas. *J. Am. Stat. Assoc.,* **86,** 278–284.

Woodruff, R.S. (1971). A simple method for approximating the variance of a complicated estimate. *J. Am. Stat. Assoc.,* **66,** 411–414.

Wright, R.L. (1983). Finite population sampling with multivariate auxiliary information, *J. Am. Stat. Assoc.,* **78,** 879–884.

Yates, F. (1933). The analysis of replicated experiments when the field results are incomplete. *Emp. J. Exp. Agric.,* **1**, 129–142.

Yates, F. and Grundy, P.M. (1953). Selection without replacement from within strata with probability proportional to size. *J. R. Stat. Soc.,* **B15**, 253–261.

Yung, W. and Rao, J.N.K. (1996). Jackknife linearization variance estimators under stratified multi-stage sampling. *Surv. Methodol.,* **22**, 23–31.

Zaslavsky, A.M. (1993). Combining census, dual-system, and evaluation study data to estimate population shares. *J. Am. Stat. Assoc.,* **83**(423), 1092–1105.

Further references

American Hospital Association, *Guide to Health Care Field* (1988).

The New York Times Almanac (1998). (Ed.) J.W. Wright

The New York Times, News Week, and *Time* Magazine: Results of public and political polls.

Statistical Abstract of the United States, 1999.

U.S. Baron's Profiles of American Colleges (1981).

Supplementary reading

Cassel, C.M., Särndal, C.E., and Wretman, J.H. (1977). *Foundations of Inference in Survey Sampling.* John Wiley & Sons, New York.

Cochran, W.G. (1997). *Sampling Techniques.* John Wiley & Sons, New York.

Deming, W.E. (1950). *Some Theory of Sampling,* Dover, New York.

Des Raj (1968). *Sampling Theory.* McGraw-Hill, New York.

Hansen, M.H., Hurwitz, W.N., and Madow, W.G. (1953). *Sample Survey Methods and Theory,* Vols. 1 and 2, John Wiley & Sons, New York.

Jessen, R.J. (1978). *Statistical Survey Techniques.* John Wiley & Sons, New York.

Kish, L. *Survey Sampling.* John Wiley & Sons, New York.

Levy, P.S. and Lemeshow, S. (1991). *Sampling of Populations: Methods and Applications.* John Wiley & Sons, New York.

Lohr, S.L. (1999). *Sampling: Design and Analysis.* Duxbury Press, New York.

Som, R.K. (1995). *Practical Sampling Techniques,* 2nd ed., Marcel Dekker, New York.

Stuart, A. (1984). *The Ideas of Sampling.* Oxford University Press, New York.

Sukhatme, P.V., Sukhatme, B.V., Sukhatme, S., and Asok, C. (1984). *Sampling Theory of Surveys with Applications,* Iowa State University Press, Ames.

Thompson, S.K. (1992). *Sampling.* John Wiley & Sons, New York.

Thompson, M.E. (1997). *Theory of Sample Surveys.* Chapman & Hall, London.

Yates, F. (1960). *Sampling Methods for Censuses and Surveys.* 3rd ed., Charles Griffin and Company, London.

Index